Recent Development in Magnetic Shape Memory Alloys

Recent Development in Magnetic Shape Memory Alloys

Editor

Volodymyr Chernenko

MDPI • Basel • Beijing • Wuhan • Barcelona • Belgrade • Manchester • Tokyo • Cluj • Tianjin

Editor
Volodymyr Chernenko
University of the Basque Country (UPV/EHU)
Spain

Editorial Office
MDPI
St. Alban-Anlage 66
4052 Basel, Switzerland

This is a reprint of articles from the Special Issue published online in the open access journal *Metals* (ISSN 2075-4701) (available at: https://www.mdpi.com/journal/metals/special_issues/magnetic_shape_memory).

For citation purposes, cite each article independently as indicated on the article page online and as indicated below:

LastName, A.A.; LastName, B.B.; LastName, C.C. Article Title. *Journal Name* **Year**, *Volume Number*, Page Range.

ISBN 978-3-0365-6762-4 (Hbk)
ISBN 978-3-0365-6763-1 (PDF)

Cover image courtesy of Aaron Acierno, Amir Mostafaei, Jakub Toman, Katerina Kimes, Mirko Boin, Robert C. Wimpory, Ville Laitinen, Andrey Saren, Kari Ullakko, Markus Chmielus.

© 2023 by the authors. Articles in this book are Open Access and distributed under the Creative Commons Attribution (CC BY) license, which allows users to download, copy and build upon published articles, as long as the author and publisher are properly credited, which ensures maximum dissemination and a wider impact of our publications.
The book as a whole is distributed by MDPI under the terms and conditions of the Creative Commons license CC BY-NC-ND.

Contents

About the Editor vii

Preface to "Recent Development in Magnetic Shape Memory Alloys" ix

Natalia A. Río-López, Patricia Lázpita, Daniel Salazar, Viktor I. Petrenko, Fernando Plazaola, Volodymyr Chernenko and Jose M. Porro
Neutron Scattering as a Powerful Tool to Investigate Magnetic Shape Memory Alloys: A Review
Reprinted from: *Metals* 2021, *11*, 829, doi:10.3390/met11050829 1

Michal Rameš, Vít Kopecký and Oleg Heczko
Compositional Dependence of Magnetocrystalline Anisotropy in Fe-, Co-, and Cu-Alloyed Ni-Mn-Ga
Reprinted from: *Metals* 2022, *12*, 133, doi:10.3390/met12010133 23

Lara Righi
Neutron Diffraction Study of the Martensitic Transformation of $Ni_{2.07}Mn_{0.93}Ga$ Heusler Alloy
Reprinted from: *Metals* 2021, *11*, 1749, doi:10.3390/met11111749 31

Andrew Armstrong and Peter Müllner
Actuating a Magnetic Shape Memory Element Locally with a Set of Coils
Reprinted from: *Metals* 2021, *11*, 536, doi:10.3390/met11040536 39

Vít Kopecký, Michal Rameš, Petr Veřtát, Ross H. Colman and Oleg Heczko
Full Variation of Site Substitution in Ni-Mn-Ga by Ferromagnetic Transition Metals
Reprinted from: *Metals* 2021, *11*, 850, doi:10.3390/met11060850 53

Sergey Kustov, Andrey Saren, Bruno D'Agosto, Konstantin Sapozhnikov, Vladimir Nikolaev and Kari Ullakko
Transitory Ultrasonic Absorption in "Domain Engineered" Structures of 10 M Ni-Mn-Ga Martensite
Reprinted from: *Metals* 2021, *11*, 1505, doi:10.3390/met11101505 71

Yulia Sokolovskaya, Olga Miroshkina, Danil Baigutlin, Vladimir Sokolovskiy, Mikhail Zagrebin, Vasilly Buchelnikov and Alexey T. Zayak
A Ternary Map of Ni–Mn–Ga Heusler Alloys from Ab Initio Calculations
Reprinted from: *Metals* 2021, *11*, 973, doi:10.3390/met11060973 89

Patricia Lázpita, Elena Villa, Francesca Villa and Volodymyr Chernenko
Temperature Dependent Stress–Strain Behavior and Martensite Stabilization in Magnetic Shape Memory $Ni_{51.1}Fe_{16.4}Ga_{26.3}Co_{6.2}$ Single Crystal
Reprinted from: *Metals* 2021, *11*, 920, doi:10.3390/met11060920 103

Zhen Chen, Daoyong Cong, Shilei Li, Yin Zhang, Shaohui Li, Yuxian Cao, Shengwei Li, et al.
External-Field-Induced Phase Transformation and Associated Properties in a $Ni_{50}Mn_{34}Fe_3In_{13}$ Metamagnetic Shape Memory Wire
Reprinted from: *Metals* 2021, *11*, 309, doi:10.3390/met11020309 115

J. D. Navarro-García, J. P. Camarillo-Garcia, F. Alvarado-Hernández, J. L. Sánchez Llamazares and H. Flores-Zúñiga
Elastocaloric and Magnetocaloric Effects Linked to the Martensitic Transformation in Bulk $Ni_{55}Fe_{11}Mn_7Ga_{27}$ Alloys Produced by Arc Melting and Spark Plasma Sintering
Reprinted from: *Metals* 2022, *12*, 273, doi:10.3390/met12020273 129

Aaron Acierno, Amir Mostafaei, Jakub Toman, Katerina Kimes, Mirko Boin, Robert C. Wimpory, Ville Laitinen, et al.
Characterizing Changes in Grain Growth, Mechanical Properties, and Transformation Properties in Differently Sintered and Annealed Binder-Jet 3D Printed 14M Ni–Mn–Ga Magnetic Shape Memory Alloys
Reprinted from: *Metals* **2022**, *12*, 724, doi:10.3390/met12050724 . . . . **139**

About the Editor

Volodymyr Chernenko

Volodymyr Chernenko (Prof. Dr.) Volodymyr Chernenko is an Ikerbasque Research Professor at the Basque Center for Materials, Applications and Nanostructures (BCMaterials) and the University of the Basque Country (UPV/EHU). His professional interest is focused on research and development of multifunctional magnetic shape memory materials, magnetomechanics of martensites, thin films and nanotechnologies of the ferromagnetic martensites. Since the late 1980s, he is known world-wide as one of the founders of the new research area "Ferromagnetic shape memory alloys (FSMAs)".He is the author of more than 350 original papers in ISI scientific journals, 9 book chapters (4 Elsevier, 3 Springer) and 6 review articles, with more than 10,600 citations and an h-index equal to 51 (Google Scholar). He has been a guest editor of five special topic books on FSMAs.

Preface to "Recent Development in Magnetic Shape Memory Alloys"

Heusler type magnetic shape memory alloys (MSMAs) have gained a strong academic and practical importance during recent decades due to the giant effects they exhibit, particularly magnetostrain produced by the magnetic field-induced twin boundaries motion and magneto(elasto)-caloric effects resulting from the magnetic field-induced martensitic transformation. The aim of this Special Issue is to address recent advances in improving the understanding of material–process–structure–property relationships in these materials. The Special Issue contains 11 articles—1 review and 10 research papers—written by well-known experts in the MSMA field representing scientific institutions from 8 countries. The Special Issue should be useful for researchers and engineers dealing with smart multifunctional materials.

Volodymyr Chernenko
Editor

Review

Neutron Scattering as a Powerful Tool to Investigate Magnetic Shape Memory Alloys: A Review

Natalia A. Río-López [1], Patricia Lázpita [1,2], Daniel Salazar [1], Viktor I. Petrenko [1,3], Fernando Plazaola [2], Volodymyr Chernenko [1,2,3] and Jose M. Porro [1,3,*]

1. BCMaterials, Basque Center for Materials, Applications & Nanostructures, 48940 Leioa, Spain; natalia.rio@bcmaterials.net (N.A.R.-L.); patricia.lazpita@ehu.eus (P.L.); daniel.salazar@bcmaterials.net (D.S.); viktor.petrenko@bcmaterials.net (V.I.P.); vladimir.chernenko@gmail.com (V.C.)
2. Department of Electricity and Electronics, Faculty of Science and Technology, University of the Basque Country, 48080 Bilbao, Spain; fernando.plazaola@ehu.eus
3. Ikerbasque, Basque Foundation for Science, 48009 Bilbao, Spain
* Correspondence: jm.porro@bcmaterials.net

Abstract: Magnetic shape memory alloys (MSMAs) are an interesting class of smart materials characterized by undergoing macroscopic deformations upon the application of a pertinent stimulus: temperature, stress and/or external magnetic fields. Since the deformation is rapid and contactless, these materials are being extensively investigated for a plethora of applications, such as sensors and actuators for the medical, automotive and space industries, energy harvesting and damping devices, among others. These materials also exhibit a giant magnetocaloric effect, whereby they are very promising for magnetic refrigeration. The applications in which they can be used are extremely dependent on the material properties, which are, in turn, greatly conditioned by the structure, atomic ordering and magnetism of a material. Particularly, exploring the material structure is essential in order to push forward the current application limitations of the MSMAs. Among the wide range of available characterization tools, neutron scattering techniques stand out in acquiring advanced knowledge about the structure and magnetism of these alloys. Throughout this manuscript, a comprehensive review about the characterization of MSMAs using neutron techniques is presented. Several elastic neutron scattering techniques will be explained and exemplified, covering neutron imaging techniques—such as radiography, tomography and texture diffractometry; diffraction techniques—magnetic (polarized neutron) diffraction, powder neutron diffraction and single crystal neutron diffraction, reflectometry and small angle neutron scattering. This will be complemented with a few examples where inelastic neutron scattering has been employed to obtain information about the phonon dispersion in MSMAs.

Keywords: magnetic shape memory alloys; neutron scattering; Heusler alloys

1. Introduction

Magnetic shape memory alloys (MSMAs) are an interesting class of smart materials exhibiting the martensitic transformation and/or twinning induced large macroscopic deformations upon the application of pertinent stimuli, such as temperature, stress and/or magnetic field (see, e.g., [1–4] and references therein). Since these deformations are rapid and contactless, these kinds of materials are being extensively investigated for plenty of potential applications [5,6]. The applications in which they can be used are crucially dependent of the material properties, these being greatly conditioned by their structure, atomic ordering and magnetic state [3,7]. Thus, exploring the crystal structure is essential in order to establish the current limits of these materials and their related interesting applications. As in the case of many other functional materials, magnetic shape memory alloys possess quite complex crystallographic nuclear and magnetic structures, and their determination by means of classic techniques (X-Ray Diffraction, Scanning and Transmission Electron

Citation: Río-López, N.A.; Lázpita, P.; Salazar, D.; Petrenko, V.I.; Plazaola, F.; Chernenko, V.; Porro, J.M. Neutron Scattering as a Powerful Tool to Investigate Magnetic Shape Memory Alloys: A Review. *Metals* **2021**, *11*, 829. https://doi.org/10.3390/met11050829

Academic Editors: Sergey Kustov and Thomas Niendorf

Received: 23 April 2021
Accepted: 14 May 2021
Published: 18 May 2021

Publisher's Note: MDPI stays neutral with regard to jurisdictional claims in published maps and institutional affiliations.

Copyright: © 2021 by the authors. Licensee MDPI, Basel, Switzerland. This article is an open access article distributed under the terms and conditions of the Creative Commons Attribution (CC BY) license (https://creativecommons.org/licenses/by/4.0/).

Microscopy, etc.) requires often to be complemented by more advanced characterization techniques. Among the possible solutions, neutron scattering stands out as an ideal probing tool to successfully determine the aforementioned crystal and magnetic structures [7–9]. In the present manuscript, after concise introductory notes about MSMAs and neutron scattering techniques provided below, a brief review highlighting a remarkable effectiveness of neutrons in the studies of different MSMAs is presented.

1.1. Magnetic Shape Memory Alloys, MSMAs

Active materials, defined as those which generate a controllable response as a result of an external stimulus, are usually employed as sensors and actuators for various industrial applications. In these devices, the impulses that lead to the sensing or actuating capacities consist in some type of stimulus (thermal, electrical, etc.), while the response consists either in a deformation or in another type of useful reaction. Shape Memory Alloys (SMAs) are well-known active materials characterized by a peculiar response, which consists in a thermally-induced recovering process of their original shape after suffering from a pseudoplastic deformation in the martensitic state: this is the conventional Shape Memory Effect, SME [10–12]. The so-called Martensitic Transformation (MT) is responsible for this effect. This structural first-order phase transformation is also responsible for many other interesting effects, such as the thermoelastic or the superelastic effects, among others. The MT involves a phase transition between two solid-state phases: the austenitic phase or austenite, which is a high symmetry, high temperature phase; and the martensitic phase or martensite, which is a low symmetry, low temperature phase. Due to the diffusionless character of the MT, both phases share the same chemical composition. The origin of the MT embodies a thermoelastic transformation, where a temperature change and/or the application of a mechanical stress leads to a change in the crystal structure of the alloy. This deformation implies a uniform lattice distortion characterized with an invariant distortive habit plane, resulting in the crystallographic correspondence between the martensitic and austenitic lattices. To minimize the elastic energy generated due to a strong shape change at the MT, the martensitic phase exhibits an inhomogeneous lattice invariant deformation by twinning. The crystallographically identical twin variants present in the martensite phase, initially randomly oriented when the thermally induced MT is accomplished, can be aligned upon the application of an external uniaxial stress that gives as a result the martensitic plasticity. Being a first-order transformation, the spontaneous MT is accompanied by a thermal hysteresis $\Delta T = T_A - T_M$, where T_A is the austenitic temperature characterizing the reverse MT obtained during heating, and T_M is the martensitic temperature corresponding to the forward MT obtained during cooling [10,13].

Magnetic Shape Memory Alloys (MSMAs) are mostly Heusler-type alloys exhibiting MT and a conventional SME, as well as, due to their magnetic nature, Large Magnetic Field Induced Strains (MFIS) either as a result of microstructural changes in the martensitic state or triggered by the MT [4,14]. As a consequence, a possible classification of the Heusler-type MSMAs can be established by dividing them in two groups: Ferromagnetic (FSMAs) and Metamagnetic (MetaMSMAs) Shape Memory Alloys [4,15].

Off-stoichiometric Ni_2MnGa Heusler alloys are the archetypical prototypes of FSMAs. They exhibit MT from the ferromagnetic austenite to the ferromagnetic martensite with a small change in their saturation magnetization. In the martensitic state they show a highly mobile twin microstructure with the twinning stress which can be varied down to about 0.05 MPa [16]. Due to a strong magnetoelastic coupling in FSMAs, the application of an external field close to the anisotropy field of the compound can produce equivalent uniaxial stress values of 1–3 MPa, large enough to drive the twin rearrangements resulting in the aforementioned large MFIS [4]. The schematic behavior of a FSMA depicted in Figure 1 shows an operation cycle of a single crystalline FSMA sample usually employed in commercially available devices. If the material is cooled below the MT temperature, T_M, a martensitic microstructure formed by crystallographically equivalent twin variants is obtained. The application of a magnetic field, H, increases the volume fraction of the twin variants with their easy-magnetization short *c*-axis of the tetragonal unit cell aligned with the field, hence elongating the sample in the direction perpendicular to the applied field. Subsequently, a mechanical compressive stress, *F*, can be applied in order to reset the sample shape, as shown in Figure 1. The initial shape will be recovered by raising the temperature above T_A, i.e., returning to the austenite phase.

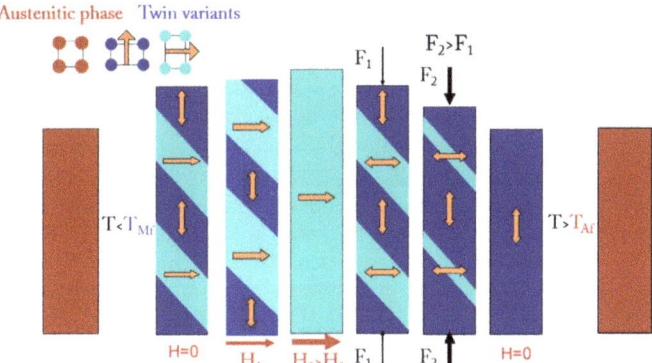

Figure 1. Schematic behavior of the process of the martensitic transformation, MT, magnetic field induced strain, MFIS, and the subsequent mechanical reset of the sample in a ferromagnetic shape memory alloy [17]. The sample is first cooled below T_M to transform from the austenite to a self-accommodated martensitic phase without applying any magnetic field ($H = 0$). The twin variants show different structural orientations and, consequently, different orientations of their magnetic moments (indicated by the orange arrows). When applying a magnetic field (H_1), the twin variants with the magnetic moments along the field grow at the expense of the others. By increasing the magnetic field (H_2), a full reorientation of the martensitic variants occurs and a single variant representing also a single magnetic domain state of the sample is reached, resulting in a macroscopic deformation. The inverse process until the former single variant state is induced by mechanical compression stress (F_1, F_2). Further heating over T_A recovers the initial shape of the sample in the austenite phase. Reproduced from Ph.D. Thesis of A. Pérez-Checa, University of the Basque Country [17].

The functionality of these compounds strongly depends on the crystal structure of the martensitic phase, the transformation behavior, the magnetic and the elastic properties of the FSMAs, which are highly sensitive to the alloy composition, lattice defects and atomic ordering. This was reconfirmed recently during the development of high temperature Ni-Mn-Ga FSMAs [17,18].

Heusler-type MetaMSMAs, represented mostly by Mn-rich Ni-Mn-X (X = Sn, In, Sb) compounds [19,20], are characterized by strong competitive ferro-antiferro magnetic exchange interactions within the unit cell of the crystal lattice. They show MT from the ferromagnetic austenite to a weak magnetic (antiferromagnetic, superparamagnetic, ferrimagnetic) martensite, which is accompanied by a huge abrupt change of the magnetization, a large specific volume change and latent heat. Contrary to FSMAs, the MT temperatures in MetaMSMAs are strongly dependent on the magnetic field enabling the magnetic field induced MT at constant temperature under the application of moderate magnetic fields. The magnetic field triggered MT gives rise to the large MFIS effect [21] or to the giant inverse magnetocaloric effect (IMCE) [22]. Whereas the study of the MFIS effect in these alloys is still at its infancy [4], the IMCE is currently subject of intensive world-wide investigations (see [15] and references therein) where, as in case of FSMAs, the composition, crystal and magnetic structures play a crucial role.

MSMAs can be utilized in several forms: single crystals, polycrystalline bulks, powders and thin films [23]. Single crystalline bulks are highly deformable in the martensitic state, but they are difficult and expensive to produce. Polycrystalline MSMAs, cheap and technologically easily accessible, show much smaller MFIS capabilities than single crystals with similar compositions, due to constrains from the grain boundaries inhibiting the twin boundary motion [24–27]. Thin films, epitaxially grown on a substrate, exhibit many more constrains from the substrate that result into difficulties for them to show uniform deformations. Nonetheless, they are promising candidates, especially in cantilever or free-standing form, for their implementation in different industrial applications, such as Micro-Electro-Mechanical Systems (MEMS) and Micro-Magneto-Mechanical Systems (MMMS) [13,28,29].

1.2. Neutron Scattering Techniques

Neutron scattering techniques are widely used in materials science due to their versatility and multidisciplinary character, being the determination of the atomic relative positions and atom mobilities in solid or liquid bulk materials one possible application example of these techniques. The basis of neutron scattering consists in the measurement of the intensity of a neutron scattered beam after the beam has passed through the sample. Neutrons possess no charge and their electric dipole is zero, one of the reasons why they can penetrate matter far deeper than charged particles. This, along with the fact that neutrons interact with atoms via nuclear rather than electrical forces, and because nuclear forces are very short ranged, allows neutrons to travel large distances through most materials without being adsorbed or scattered. The interactions between neutrons and atomic nuclei allow them to differentiate isotopes and nearby elements in the periodic table, which is an advantage over other techniques such as X-ray or electron diffraction. Neutrons can probe not only the nuclear structure of materials, but also their magnetic properties. This is a consequence of the fact that neutrons possess a net magnetic moment, so they interact magnetically with the electrons in the sample, resulting in magnetic scattering events whose analysis yields specific information about the magnetic structure of the sample [17,30].

Especially in the case of the Heusler type Ni-X-Y (X,Y = Mn, Ga, Sn, In, Sb, Fe) MSMAs, neutron scattering has a much better precision in the determination of chemical ordering compared to X-ray diffraction. This is due to the very different neutron scattering lengths between elements that have similar Z number, as they are independent of the atomic number Z [31]. The differences in the scattering lengths between neutron and X-ray scattering events in elements commonly employed in MSMAs are shown schematically in Figure 2.

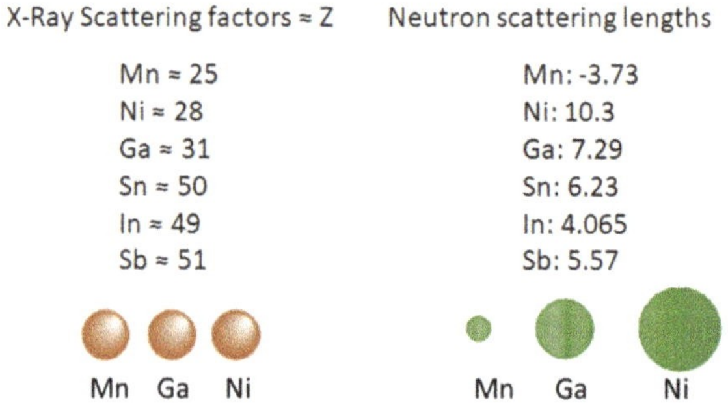

X-Ray Scattering factors ≈ Z   Neutron scattering lengths

Mn ≈ 25   Mn: -3.73
Ni ≈ 28   Ni: 10.3
Ga ≈ 31   Ga: 7.29
Sn ≈ 50   Sn: 6.23
In ≈ 49   In: 4.065
Sb ≈ 51   Sb: 5.57

Mn Ga Ni   Mn Ga Ni

Figure 2. X-ray scattering form factors and neutron scattering lengths of several elements commonly found in Heusler type alloys. A visual comparison of these is shown for Ni-Mn-Ga FSMA.

Neutron scattering techniques can be classified in two large groups: elastic and inelastic, usually employed to probe the structure and the lattice dynamics, respectively, of the materials being investigated. The former group does not imply energy exchange between the neutrons and atoms in the sample, while the latter does. Even if inelastic neutron scattering remains an almost unexplored terrain to study MSMAs, this review presents a few studies where these techniques have been employed in these materials, together with a comprehensive recompilation of studies that make use of elastic neutron scattering to probe MSMAs.

Among the elastic neutron scattering techniques, three subgroups appear: neutron imaging (radiography/tomography and texture diffractometry), large scale structures (reflectometry and small-angle scattering) and diffraction (both in powder and in single crystal), all of them including polarized (magnetic structure) and non-polarized (nuclear structure) neutron scattering experiments. As already mentioned, MSMAs usually present complex crystallographic nuclear and magnetic structures, so their properties are greatly conditioned by the crystal structure and atomic ordering. As an example, two alloys with the same composition can exhibit different transformation and magnetic characteristics if the atoms in the unit cell are arranged in a different way. In order to predict how a specific alloy will behave, a profound understanding of the exact atomic ordering is crucial, an aspect in which neutron scattering techniques excel in MSMAs. In this framework, several situations in which each of the aforementioned techniques have been employed to characterize MSMAs are reviewed hereafter [17,30,32,33].

Neutron scattering techniques provide scientists investigating MSMAs with unique tools to access information in these alloys that cannot be accessed anyhow else, as it is the case of obtaining not only the atomic site occupancies, but also the magnetic site densities with majestic precision. Moreover, the presence of inhomogeneities in the bulk samples, both structural and magnetic, can uniquely be determined by neutron scattering and cannot be explored by any other technique. These unique capabilities demonstrated by neutron scattering techniques compensate the efforts needed to access neutron sources, where neutron beam times are awarded to scientists upon the presentation of a beam time proposal, which is evaluated by a panel of experts and by the local scientists of the neutron sources.

2. Elastic Neutron Scattering Studies of MSMAs

2.1. Diffraction

2.1.1. Powder Neutron Diffraction

Powder diffraction is the neutron scattering technique employed when the crystal structure of powder samples needs to be explored [34,35]. It is used to detect and identify crystalline phases, to quantitatively determine their lattice parameters and volume fractions and, in general, to characterize atomic arrangements and the microstructure of polycrystalline materials in powder form. There are two types of powder neutron diffraction methods: angle-dispersive and energy-dispersive. In the conventional method of neutron powder diffraction, also known as angle-dispersive or fixed-wavelength, neutrons of a fixed wavelength are selected by a crystal monochromator. These neutrons are then scattered by the sample and the intensity $I(\theta)$ of the scattered beam is measured as a function of the scattering angle 2θ. An experimental plot of $I(\theta)$ vs. θ shows diffractions peaks whose positions are determined by the Bragg law $n\lambda = 2d\sin\theta$, where d is the interplanar spacing between the different lattice planes of the crystal structure, n is an integer number and λ is the neutron wavelength. By reversing the roles of θ and λ, it is possible to measure the intensity $I(\lambda)$ vs. λ at a fixed value of 2θ. This is the fixed-angle or energy-dispersive method. In the first method, the "white" (non-monochromatic) neutron beam from the reactor passes through a collimator in the reactor shield, and a particular wavelength is then selected by a crystal monochromator, as previously mentioned. In the latter, the wavelength of the neutrons is determined by measuring the time it takes for them to reach the detector: this is the so-called time-of-flight neutron scattering method [32,33,36,37].

Neutron powder diffraction has historically been used to study Ni-Mn-Ga FSMAs. Back in the early 1980s, Webster et al. [38] employed neutron powder diffraction to study the structure and the structural phase transformations of a stoichiometric Ni_2MnGa Heusler alloy. These measurements played an important role in confirming that the crystal structure is highly $L2_1$-ordered, schematically represented in Figure 3, and that the structural transformation turns out to be reversible during cooling-heating cycles. In the early 2000s, the growing interest in these alloys led to the occurrence of several neutron powder diffraction studies in off-stoichiometric and doped Ni-Mn-Ga FSMAs. High resolution powder neutron diffraction was used by Brown et al. [39] to study the martensitic and premartensitic transformations in Ni_2MnGa, as well as the distribution of Ni and Mn atoms within the unit cell. Moreover, the ability of neutrons to distinguish between Ni and Mn allowed to observe the temperature-induced atomic displacements in each sublattice. Cong et al. [40] found that the transformation process in $Ni_{53}Mn_{25}Ga_{22}$ alloys is different from that investigated in stoichiometric Ni_2MnGa alloys, since a pretransformation mechanism seemingly starts in the martensitic phase rather than in the austenitic one. They also demonstrated that the martensitic phase presents a non-modulated tetragonal structure. The same research group [41] analyzed three particular FSMAs ($Ni_{53}Mn_{25}Ga_{22}$, $Ni_{48}Mn_{25}Ga_{22}Co_5$ and $Ni_{48}Mn_{30}Ga_{22}$) by neutron powder diffraction in order to observe how doping with Co and changing the alloy composition affect the crystal structure at room temperature. The last composition of the three alloys studied presents a cubic austenitic structure at room temperature, whereas the other two present the same martensitic tetragonal structure. They also confirmed that the substitution of Ni for Co does not affect the crystal structure, whereas substituting Ni for Mn does. Orlandi et al. [42,43] investigated the influence of Co doping on the martensitic transformation in stoichiometric Ni_2MnGa. They found that, whereas in Ni_2MnGa and $Ni_{47}Co_3Mn_{25}Ga_{25}$ the MT takes place in two steps and involves a premartensitic phase, in $Ni_{45}Co_5Mn_{25}Ga_{25}$ the transition occurs in a single step. They also reported on the presence of long-range antiferromagnetic ordering in the Co-rich Ni-Co-Mn-Ga alloy.

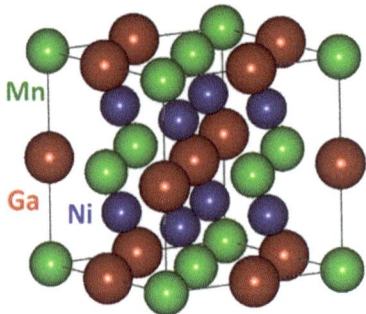

Figure 3. Crystal structure of the Heusler Ni_2MnGa alloy in the austenitic $L2_1$ cubic phase.

As previously stated, powder neutron diffraction is an excellent tool to analyze atomic ordering. Richard et al. [44] took advantage of this method to study a set of Ga-deficient Ni-Mn-Ga alloys, with excessive Mn and 48–52 at.% Ni. They asserted that, when Ni is presented with a 50 at.%, the Mn in excess occupies Ga sites. When the Ni content is below 50%, Mn in excess tends to occupy Ni and Ga sites, whereas for Ni contents above 50 at.%, excess Ni atoms occupy Mn sites by displacing Mn to Ga sites. As a result of these atomic orderings, Mn atoms at the Ga sites become nearest neighbors to Mn atoms in their proper sites, leading to antiferromagnetic coupling and, hence, to a reduction in the total magnetic moment. These studies were completed by Lázpita et al. [45,46] who recompiled all the data presented by Richard et al. adding a new composition. They studied the martensitic structures and cell parameters of a set of Ga-deficient Ni-Mn-Ga alloys with excess Mn and 43–52 at.% Ni. Samples with excess Ni show a 5M tetragonal martensite structure, while defective Ni alloys present a 7M-modulated orthorhombic structure with a small monoclinic distortion. The nature of the magnetic coupling between Mn atoms at different sites was analyzed in terms of the dependence of the sign of the magnetic exchange integral on the distances between Mn atoms.

Neutron diffraction was used by Wang et al. [47] to disclose the existence of amorphous phases and a MT from tetragonal to cubic phases in the crystallized fraction of $Ni_{57}Mn_{27}Ga_{22}$ nanoparticles prepared by ball-milling and subsequent post-annealing. The amorphous phase in this case was considered to control the transformation kinetics.

Neutron powder diffraction was also employed for the analysis of the atomic site occupancies in more complex doped Ni-Mn-Ga alloys. Roy et al. [48] studied the effect of doping on the properties of Ni_2MnGa. Since Cu tends to occupy Mn sites, Mn atoms are pushed to occupy Ga sites. As a consequence, the Curie temperature (T_C) is lowered and T_M raised, so that, at 25 at.% of Cu the martensitic and magnetic transitions in $Ni_2Mn_{0.75}Cu_{0.25}Ga$ coincide at 317 K. Pérez-Checa et al. [18] examined by neutrons a set of six Ni-Mn-Ga-Co-Cu-Fe high temperature FSMAs to disclose the influence of Fe doping on the evolution of atoms distribution within the unit cell of the crystal lattice and to correlate the Fe-triggered atomic redistribution with their magnetic properties. They found that the Fe increase provokes an enhancement of the ferromagnetically coupled Mn-Mn pairs at the expense of the antiferromagnetic ones, leading to an increase of the total magnetic moment of the alloy, in agreement with magnetometry measurements on the alloys.

Neutron studies on Ni-Mn-based MetaMSMAs started in the late 2000s. Brown et al. [49] investigated $Ni_2Mn_{1.94}Sn_{0.56}$ by neutron powder diffraction in order to determine the atomic positions within the unit cell and the crystal structure of the martensitic phase. They concluded that Mn atoms in excess occupy vacant Sn sites, and that the martensitic structure is a 4M-modulated orthorhombic one. Mañosa et al. [50] used neutron diffraction to reveal the magnetic field-induced martensitic transformation in the $Ni_{49.7}Mn_{34.3}In_{16.0}$ MetaMSMA, associated with the strong coupling of magnetism and structure. They showed that this coupling led to magnetic superelasticity, magnetic shape memory, giant magnetocaloric and giant inverse magnetocaloric effects. Brown et al. [51] studied a Mn_2NiGa alloy with powder neutron diffraction, revealing that in the parent cubic phase (000) sites are occupied by Mn, (1/2,1/2,1/2) sites by Ga and (1/4,1/4,1/4) and (3/4,3/4,3/4) sites by a mixture of Ni and Mn. Furthermore, from the recorded diffraction patterns at 5 K they deduced that the coupling between Mn atoms in the martensite phase in this alloy is of ferromagnetic nature. Mukadam et al. [52] employed neutron powder diffraction and complementary magnetic characterization techniques in order to study how an increase in Ni and a decrease in Mn contents affect the magnetism in a set of $Ni_{2+x}Mn_{1-x}Sn$ alloys. They concluded that an increase in the Ni content at the expense of Mn atoms produced a reduction of both T_C and the total magnetic moment of the unit cell.

2.1.2. Single Crystal Neutron Diffraction

Single crystal neutron diffraction measures the coherent scattering intensities (Bragg intensities) from a single crystal, so that the crystal structure of the material can be analyzed [53]. The unit cell, space group, positions of the atomic nuclei and site occupancies can be determined with this technique. Since neutrons possess a magnetic moment, the magnetic structure of the material can be determined by analyzing the magnetic contributions to the Bragg peaks. When a single crystal is placed in a beam of neutrons, each set of planes in the sample diffract in certain distinct directions. In order to determine every plane set of the crystalline structure, either an area detector must be employed, or the sample has to be rotated around one crystallographic axis when using a point detector. Figure 4 shows the differences in the diffraction conditions and the signal recording between polycrystalline and single crystal samples. The atomic interplanar spacings can be determined by knowing the measurement angles and employing the Bragg's law [54–57]. For both single crystal and powder neutron diffraction, and by making the corresponding mathematical corrections to the recorded intensity profile at the detector, it is possible to refine the structure and atomic ordering from fittings to that intensity profile using structure refinement programs such as, e.g., Fullprof [58].

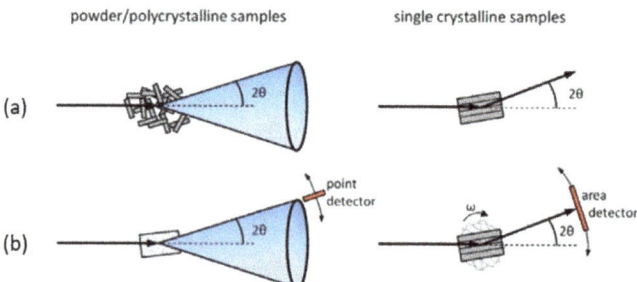

Figure 4. Schematic of sample type (**a**) and experimental setup (**b**) for powder/polycrystalline sample (left column) and single crystalline samples (right column). The incoming beam is diffracted taking the shape of a cone both in powder and polycrystalline samples, therefore a point detector being enough to record all the reflections from the sample. Nonetheless, single crystalline samples diffract in specific directions, and the sample must be aligned for each 2θ position of the area detector, being the sample also rotated around its ω axis. Reproduced from Ph.D. Thesis of M. Chmielus, Technische Universität Berlin [55].

Historically, single crystal neutron diffraction has been employed first to determine the structure and atomic positions in ferromagnetic shape memory alloys. Brown et al. [39] carried out single crystal and powder neutron diffraction experiments to establish the diagram of the structural phase transitions in alloy compositions close to the stoichiometric Ni_2MnGa. Glavatskyy et al. [59] investigated the effect of alloying on the crystal structure, lattice parameters, MT temperatures and magnetoplasticity of single crystalline $Ni_{49.4}Mn_{23.3}Ga_{25.6}Cu_{1.7}$, $Ni_{49.3}Mn_{27.8}Ga_{20.9}Cu_{2.0}$, $Ni_{47.3}Mn_{25.5}Ga_{24.5}Cu_{2.7}$, $Ni_{46.8}Mn_{27.3}Ga_{22.8}Cu_{3.1}$ and $Ni_{49.6}Mn_{27.6}Ga_{18.5}Cu_{3.9}$ FSMAs. Neutron diffraction measurements revealed a change in the structure of the martensitic phase from a 5 M to a 7 M modulated body-centered tetragonal (bct) as a function of the Cu content. Alloying with Cu affected both T_C and T_M due to changes in the valence electron concentrations and in the Mn-Mn and Mn-Ni exchange interactions. Brown et al. [60] examined the crystal structure at different temperatures, the transition temperatures (T_C and T_M) and the MT mechanism under applied stress of a Ni-Fe-Ga FSMA. They found that a cubic-to-tetragonal MT occurs without any orthorhombic intermediate phase in these alloys. Single crystal and powder neutron diffraction techniques were combined by Richard et al. [44] and Lázpita et al. [45,46] to disclose the atomic positions in a set of Ni-Mn-Ga alloys (specific details of their findings are found in the powder neutron diffraction section of the present manuscript). Both types of diffraction in these works confirmed the existence of similar crystal structures of the martensitic phase, as well as similar atomic site occupancies and transition temperatures both in single crystalline and in powder forms.

The transformation behavior of the $Ni_{50.5}Mn_{28.2}Ga_{21.2}$ FSMA single crystal was studied by neutron diffraction in the temperature range of 4–300 K by Glavatskyy et al. [61]. They found that an orthorhombic martensitic structure with 5 M modulation remains stable in the 4–300 K temperature range, while the martensite twin mobility suddenly rises at 200 K. Molnar et al. [62] explored by neutron diffraction a stress-induced martensitic variant reorientation in a $Ni_{49.7}Mn_{29.3}Ga_{21}$ single crystal exhibiting a 5 M tetragonal martensite. They demonstrated that the macroscopic strain originates only from the variant redistribution, since no other contributions to that strain were detected. Chmielus et al. [63–65] studied the presence of different martensitic variants in Ni-Mn-Ga FSMAs in comparable ($Ni_{50.6}Mn_{28.3}Ga_{21}$) or distinct ($Ni_{50.5}Mn_{28.7}Ga_{20.8}$) volume fraction alloys. They also determined the lattice parameters and modulation periods in the martensitic phase in these alloys. Kabra et al. [66] studied a Ni_2MnGa alloy by single crystal neutron diffraction to determine the crystallographic orientation relationship between the twinned and untwinned regions. They demonstrated that the orientation relation between both twin variants is a 90° rotation around the b-axis of the original crystal.

2.1.3. Polarized Neutron Diffraction

Neutrons are electrically uncharged but possess a magnetic dipole moment. This moment interacts with the magnetic field of unpaired electrons in the 3d orbitals of the MSMA sample, through either the magnetic field associated with the orbital motion of the electron or the intrinsic dipole moment of the electron itself. As a result, magnetic scattering events occur in addition to the already mentioned structural scattering from the atomic nuclei. When neutrons have their magnetic moments oriented randomly (unpolarized neutron beam) they interact magnetically with the sample and are scattered in all directions, so that the net contribution to the total scattering cross section is usually very low. However, in a polarized neutron beam all the neutron magnetic moments are pointing towards the same direction and the pure magnetic scattering cross section becomes relevant. To polarize and monochromatize the neutron beam, the white unpolarized beam goes through a polarized single crystal which acts both as monochromator and also as spin filter (see Figure 5). Once the beam is polarized, the neutron beam passes through a spin flipper which modifies the orientation of the polarization, giving rise to neutrons with spins up or down. The analysis of the magnetic moment of the scattered polarized neutrons after hitting the sample allows to obtain a pure magnetic diffraction profile. As a result, the magnetic structure of the sample can be determined. The main difference between the theory of magnetic scattering of neutrons and the theory of nuclear scattering is that magnetic scattering is highly dependent on the direction of the applied magnetic field, leading to a magnetic cross-section with a vector component, whereas the cross-section of nuclear scattering is a scalar quantity [32,33,36,67].

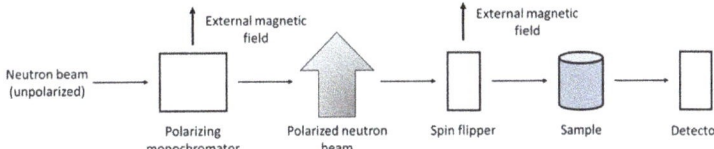

Figure 5. Scheme of a polarized neutron diffraction experimental set-up.

Polarized neutron diffraction is usually employed in single crystal alloys. Since the early 80s, the magnetic structure of Ni-Mn-Ga based FSMAs studied by polarized neutron diffraction has been the focus of several works. Webster el al. [38] employed it to demonstrate that the magnetic moments in the Ni_2MnGa alloy are associated mainly to the Mn sites, with a small contribution from the Ni sites. Brown et al. [68] analyzed the MT in a Ni_2MnGa alloy, observing a magnetic moment transfer from Mn to Ni. This redistribution of electrons between 3d sub-bands of different symmetries led to the conclusion that the cubic-to-tetragonal phase transition is driven by a band Jahn-Teller distortion. Cong et al. [40] complemented neutron powder diffraction studies on $Ni_{53}Mn_{25}Ga_{22}$ by performing polarized neutron diffraction experiments at different temperatures. They concluded that the magnetization becomes weaker as the temperature increases due to the decreasing magnetic ordering, and that there is no intermartensitic transformation in the investigated alloy. Pramanick et al. [69] employed polarized neutron diffraction to establish a correlation between the rotation of magnetic moments and the twin-reorientation phenomena in a $Ni_2Mn_{1.14}Ga_{0.86}$ single crystal. Lázpita et al. [70] performed polarized neutron diffraction measurements to determine the influence of the atomic positions within the crystal unit cell on the magnetic coupling between atoms in the austenitic phase of $Ni_{51.9}Mn_{26.2}Ga_{21.9}$, and in the martensitic phase of $Ni_{52.6}Mn_{26.9}Ga_{20.5}$, the latter presenting a non-modulated tetragonal phase. The analysis of the spin density maps (Figure 6) demonstrates that the main differences between both alloys are the Mn-Mn interatomic distances, which modify the electronic structure and significantly change the total magnetic moment of the alloys. In the austenite phase all the Mn atoms on Ga sites are coupled antiferromagnetically with Mn atoms in their own sites, whereas in the martensitic phase

Mn atoms in these sites couple both ferro- and antiferromagnetically depending on their neighboring atoms. With these results, they demonstrated a crossover of the Mn-Mn coupling, from ferromagnetic to antiferromagnetic, when the Mn-Mn atomic distances change from 3.32 Å to 2.92 Å, respectively.

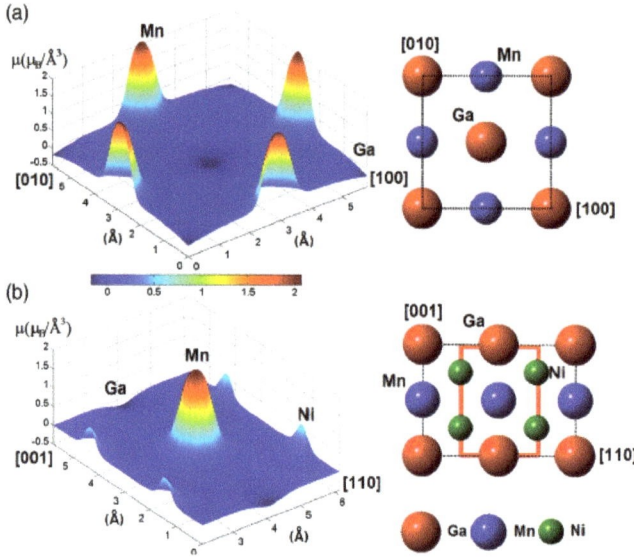

Figure 6. Spin density maps for a $Ni_{52}Mn_{26}Ga_{22}$ FSMA single crystal. These maps were obtained from single crystal polarized neutron diffraction experiments carried out at the D3 instrument (ILL, Grenoble). The maps show the spin density distribution in the (001) plane (**a**) and in the (−110) plane (**b**) of the $L2_1$ structure measured in the ferromagnetic austenite at 330 K under 9 T. Mn and Ni sites show a positive spin density that corresponds to the ferromagnetic coupling of these atoms, while Ga sites present a negative contribution attributed to the antiferromagnetically coupled Mn. The right side of the figure shows the atomic distribution in the aforementioned planes obtained by means of non-polarized single crystal neutron diffraction experiments (D9, ILL). The rectangular red line indicates the fragment of the (−110) plane represented in the corresponding map [70]. Reprinted Figure 4 with permission from P. Lázpita et al., Physical Review Letters 119, 155701 (2017), Copyright (2017) by the American Physical Society.

Neutron diffraction has been extensively used to determine not only the effect of atomic site occupancies on the properties of MSMAs, but also the characteristics of the martensitic transformation, being a valuable tool to identify premartensitic phases in these alloys. The criticality of small stoichiometric variations on the MSMA compositions has been identified by several authors, finding that variations as small as 1% in some of the elements forming the MSMA can lead to dramatic changes in the MT temperatures (T_M, T_A), as well as in the magnetic ordering ones (T_C). Moreover, the role of different dopants on the properties of MSMAs has been extensive issue of study by means of neutron diffraction. In this way, the effect doping elements have on the crystalline structure of the martensite phases, leading not only to the appearance of crystal structure modulations but also to a change on the crystal lattice space group, has been addressed by many groups studying MSMAs. Besides affecting the crystal lattices of MSMAs, dopants also have a strong influence on the martensite transformation and magnetic ordering temperatures. Finally, polarized neutron diffraction allows studying the magnetic spin densities in each specific site of the crystal lattices of the MSMAs. This opens the door to investigate the magnetic coupling regimes between magnetic elements within the alloy, and changes on the magnetic properties of the MSMAs upon on-purpose stoichiometry modifications and/or the addition of doping elements.

2.2. Large Scale Structures

2.2.1. Small-Angle Neutron Scattering (SANS)

Small-Angle Neutron Scattering (SANS) is a coherent elastic neutron scattering technique largely used for the study of nanoscale inhomogeneities in materials, involving neutron wavelengths typically in the range of tens to hundreds of Angstroms [71]. Since the characteristic sizes studied by SANS are large compared to interatomic distances, the matrix (sample medium) in which these inhomogeneities are embedded is usually treated as continuum and characterized by an average scattering length density. The presence of these inhomogeneities, such as dispersed particles or, as it is the case in MSMAs, different crystallographic phases and/or compositions, is detected thanks to the variation of the scattering length density of these nano-inhomogeneities with respect to the medium (homogeneous matrix) [33,36,72,73].

SANS plays an important role in the study of phase-separated MSMAs. This is the case of the studies performed by Runov et al. [74,75], who employed Small-Angle Polarized Neutron Scattering (SAPNS) to study a $Ni_{49.1}Mn_{29.4}Ga_{21.5}$ single crystal at different temperatures and under magnetic fields. They found an asymmetry in the polarized neutron scattering intensity profiles at around 150 K, which led them to determine the coexistence of two different phase transformations: the martensitic phase transformation expected at 306 K and an unexpected one at 150 K. The different magnetic orderings present in each phase are responsible for the change in polarization observed with SAPNS. Using standard (non-polarized) SANS, an interesting work has been performed by Sun et al. [76,77], who analyzed the decomposition of a face centered cubic (fcc) martensite phase in $Mn_{19}Cu_2Al_4Ni$, accompanied by the formation of Cu-rich clusters and an α-Mn phase under different ageing treatments. They observed fewer and larger clusters when higher temperature and longer time ageing processes were used. They also determined that the martensitic phase decomposition is a process characterized by a two-step linear kinetic regime: the first one with a larger slope was observed for aging times below 16 h, and the second one with a smaller slope for aging times larger than 16 h. The presence of ferromagnetic nanoprecipitates in an antiferromagnetic matrix was reported by Benacchio et al. [78] in a bulk $Ni_{50}Mn_{45}In_5$ MetaMSMA field-annealed at 700 K. The results were obtained by combining unpolarized and spin-polarized SANS. This phenomenon was also studied by Sarkar et al. [79], who demonstrated the presence of spin-clusters of structural origin in a $Ni_{45}Co_5Mn_{38}Sn_{12}$ MetaMSMA, which are related to the martensitic transformation. Kopitsa et al. [80] and Bliznuk et al. [81] studied the nuclear and magnetic structures in FeMn-based alloys doped with Si, Cr or Ni, affected by interstitial C and N atoms. Fe-Mn, Fe-Mn-Si, Fe-Mn-Si-Cr, Fe-Mn-Si-Cr-Ni solid solutions were analyzed, revealing that Si enhances the chemical homogeneity of the studied alloys, N does not influence substantially their homogeneity and C worsens it.

2.2.2. Reflectometry

This technique is used in MSMA thin films, commonly in combination with XRD and neutron diffraction techniques, and consists in the study of the characteristics of a neutron beam reflected from the sample. The intensity profile of the reflected beam is recorded as a function of the incident and reflected angles, or of the neutron wavelengths. The shape of the reflectivity profile yields information about the thickness, density and roughness of any thin film (with thickness from several nm to several hundreds of nm) layered onto the substrate [82]. The presence of periodic maxima and minima in the intensity profile determines the thickness of the layer or layers composing the sample, being the number of superimposed maxima and minima directly related with the number of layers present in the film. The surface roughness of the film is determined by the shape of the intensity curve, which falls faster for rougher samples. The one-dimensional scattering length density (SLD) profile of the sample along its surface normal, with nanometer resolution, is derived from the analysis of the reflectometry data. The main advantage of this technique lies in the possibility of studying textures and interplanar distances, alongside with the fact that neutron reflectometry can differentiate if layers are made of different metals [83].

If the neutron beam is polarized, it is possible to record polarized neutron reflectometry curves that give information about the magnetic depth profile of the sample being investigated. Normally four reflectivity curves are obtained during experiments with polarized neutrons. Spin-flip (R+ − and R− +; changes in the polarization of the neutrons after reflection) and non-spin-flip (R+ + and R− −; neutrons that keep the initial polarization after reflection) reflectivity processes are usually analyzed. The "+" is used when the spins of the neutrons are parallel to the applied external magnetic field, the "−" is used when the spins are antiparallel to the field. It should be mentioned that in-plane magnetization components which are not parallel to an external magnetic field contribute to spin-flip process (spin-down neutron will be reflected as a spin-up neutron (R− +) and vice-versa). The difference of R+ + and R− − clearly indicates the presence of an in-plane magnetic moment (magnetic SLD components) collinear to the external field. In this way, reflectometry allows to determine how magnetic domains are distributed through the depth of the thin film. Hence, polarized neutron reflectometry is a very versatile technique to study surfaces, interfaces and layer compositions in multi-layered thin films, both structurally and magnetically [84–86]. Even if this is not an outstanding technique for the study of bulk and powder MSMAs, it can provide useful information about MSMA thin films. Granovsky et al. [87] performed polarized neutron reflectometry measurements to analyze the induced magnetic moment in a 25 nm-thick $Ni_{50}Mn_{35}In_{15}$ thin film. Under an applied magnetic field of 5 kOe, results show no detectable in-plane magnetization at room temperature, while an induced magnetic moment collinear with the applied magnetic field is present at low temperatures.

Although not as extensively as neutron diffraction, small angle neutron scattering and neutron reflectometry have also been employed to investigate MSMAs by using polarized neutron beams. In particular polarized SANS has been employed to investigate the magnetic orderings of martensite phases in MSMAs, as well as the presence of magnetic spin clusters in these alloys upon doping and/or the use of interstitial light elements to enhance the MSMA properties. Polarized neutron reflectometry has been employed to investigate the magnetic depth profile of MSMA thin films with and without the presence of an external magnetic field.

2.3. Neutron Imaging

2.3.1. Texture Diffractometry

Texture diffractometry measurements are used to determine the orientation distribution of crystalline grains within a polycrystalline sample. A material is considered to be textured if the grains are crystallographically aligned along a preferred direction. This neutron imaging method for texture analysis is based on neutron diffraction, similarly to its analogue X-ray diffraction (XRD) technique [88]. Texture diffractometry measurements give as a result images with the textures present in the material. As already mentioned, the main advantage of neutron diffraction over X-ray diffraction arises from the fact that the interaction of neutrons with the material is relatively weak and it is not related to the number of protons (Z) of the elements composing the sample, and that the penetration depth of neutrons is much larger than the one of X-rays [89–91]. The application of this technique offers distinct advantages in texture determinations, particularly for samples with low angle reflections where intensity corrections for X-rays are most critical. As previously discussed in other neutron scattering techniques, it is possible to measure magnetic scattering events that give, as a result, magnetic texture analysis. This is particularly interesting for crystals with an antiferromagnetic component, where peaks exist in the diffraction pattern which are due solely to the magnetic scattering [89–91].

Neutron diffractometry imaging has been used for texture analysis in polycrystalline Ni-Mn-Ga FSMAs by several authors. Cong et al. [41], alongside with powder diffraction studies, performed texture measurements in $Ni_{53}Mn_{25}Ga_{22}$ and Co-doped $Ni_{48}Mn_{25}Ga_{22}Co_5$ alloys, concluding that the presence of strong textures is due to the hot-forging processing of the alloys. Several texture changes caused by the rearrangement of martensitic variants were observed in the Co-doped alloy during deformation, concluding as a consequence that changes in texture are closely related to the shape memory effect. The changes in texture on $Ni_{48}Mn_{30}Ga_{22}$ and $Ni_{53}Mn_{25}Ga_{22}$ FSMAs were studied by Nie et al. [92], who observed that after a compression stress was imposed on the parent phase, a strong preferred selection of two martensitic twin variants was observed in the obtained martensitic phase. Chulist et al. [93] studied the impact of the fabrication process on the texture presence in the alloys, analyzing two polycrystalline samples fabricated by directional solidification ($Ni_{50}Mn_{29}Ga_{21}$) and hot rolling ($Ni_{50}Mn_{30}Ga_{20}$). The solidified alloy is characterized by <100> fiber textures along the growth direction, while the hot rolling processing gave rise to a weak {111} <112> recrystallization texture in the alloy.

2.3.2. Radiography/Tomography

In radiographic methods, the attenuation of an incident neutron beam on passing through an object is used to study the internal structure (at micrometer length scale) of this object without destroying it. By using this radiography technique, 2D images of the sample being investigated are obtained, while by using tomography techniques, 3D images are obtained. Neutron radiography and tomography techniques are preferable over equivalent X-Rays techniques due to the high penetration depth of neutrons, which allows large samples to be investigated. Moreover, light elements can be detected in an environment dominated by heavy elements, as occurs in every neutron technique. Among the advantages of these methods, the possibility of obtaining maps of grain shapes and crystallographic orientations are noteworthy [94,95].

Kabra et al. [66] complemented single crystal neutron diffraction studies with energy dispersed neutron imaging, yielding as a result the morphology of the twinned regions in a stoichiometric Ni_2MnGa alloy. They found that this single crystal can spontaneously twin upon the application of an external field, showing the orientation relation between the untwinned and twinned regions, together with their specific morphology and the mosaic microstructure of the original crystal. Samothrakitis et al. [96] investigated the 3D microstructure of a hot-extruded $Co_{49}Ni_{21}Ga_{30}$ FSMA by neutron diffraction tomography. They found that no preferred crystallographic orientations can be appreciated in the nearly spherical grains observed in the sample.

The use of neutron imaging techniques in MSMAs has helped understanding the origin of the presence of textures in these alloys, attributing it to either the presence of doping elements, off-stoichiometric compositions or the alloy casting technique employed.

3. Inelastic Neutron Scattering Studies of MSMAs

Inelastic neutron scattering (INS) techniques are usually employed with the aim of studying the disorder in the crystalline structure of MSMAs, either positional or substitutional [97]. Due to this disorder, the intensity of the typically obtained Bragg spots within a neutron diffraction technique is reduced, as a consequence of the fact that the whole crystal does not contribute coherently to the diffraction spots due to lattice imperfections. Otherwise, in a perfect sample with a completely uniform crystal lattices the diffuse scattering (which is of inelastic nature) would be zero, and the Bragg spots would be sharp. Nonetheless, the positions of the Bragg spots are not affected. The intensity changes due to diffuse scattering events are redistributed along the diffractogram. The distribution of these intensity changes may be either uniform across the whole reciprocal lattice or concentrated in a particular anisotropic direction according to the nature of disorder correlations between neighboring cells. This diffuse scattering, thus, arises from the local configuration of the material: it is a short-range effect, while the long-range structural order does not contribute to it [33,36,98,99].

As stated in the introduction to this review, diffuse scattering is not widely used for this kind of alloys. Nevertheless, there are a few cases where this technique has been employed to study phonon dispersion in MSMAs. Zheludev et al. [100,101] studied the phonon spectra in a Ni_2MnGa single crystal in a wide temperature range above 220 K, which corresponds to the temperatures of stability of the cubic phase. An incomplete, but strong, softening at wave vector $\xi_0 \approx 0.33$ in the $[\xi\xi 0]$ TA_2 phonon branch was observed both above and below T_C of the austenite phase. During cooling, the frequency of this ξ_0 soft mode was strongly reduced down to zero at about $T_I = 250$ K which resulted from a premartensitic first-order phase transition from austenite into a soft-mode condensed intermediate phase. The origin of the phonon anomaly was attributed to the electron-phonon interactions. Similar inelastic neutron scattering measurements were performed in Ni_2MnGa by Recarte et al. [102], who investigated the influence of the magnetic field on the TA_2-phonon branch in the temperature range where the aforementioned intermediate transition takes place. As a result, a strong enhancement of the magnetoelastic interactions at the intermediate transition and the significant role of the magnetism in the phonon dispersion events were revealed.

Even when most neutron diffuse scattering experiments were focused on Ni-Mn-Ga alloys, other NiMn-based alloys were also investigated by this technique. Zheludev et al. [101] and Moya et al. [103,104] concluded that the results of the diffuse polarized neutron scattering carried out in several experiments in Ni-Mn-Ga, Ni-Mn-Sn, Ni-Mn-Sb, Ni-Mn-Al and Ni-Mn-In (Figure 7) alloys demonstrate unequivocally that, similarly to what occurs in body-centered cubic (bcc) alloys, acoustic phonons in the transverse TA_2 branch possess energies significantly lower than those in the other branches. Furthermore, they also demonstrated that phonons along other symmetry directions also possess energies larger than those of the TA_2 branch.

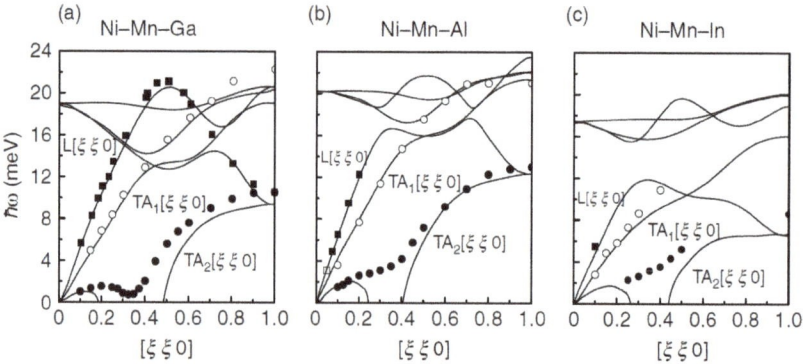

Figure 7. Phonon dispersion curves along the $[\xi, \xi, 0]$ direction for Ni_2MnGa (**a**), $Ni_{54}Mn_{23}Al_{23}$ (**b**) and $Ni_{49.3}Mn_{34.2}In_{16.5}$ (**c**). Symbols correspond to inelastic neutron scattering data (diffuse scattering) at temperatures well above the martensitic transition; lines correspond to the ab-initio calculations of the phonon dispersion curves along the experimentally determined directions in the stoichiometric Ni_2MnZ [2]. Reprinted from Handbook of Magnetic Materials, 19C, M.Acet, Ll.Mañosa and A.Planes, Magnetic-Field-Induced Effects in Martensitic Heusler-Based Magnetic Shape Memory Alloys, 231–289, Copyright (2011), with permission from Elsevier.

Inelastic neutron scattering experiments on MSMAs reveal the changes on their phonon dispersion curves when the alloys suffer from either a structural or a magnetic phase transition. This allows to identify unambiguously the phonon dispersion curves with the specific structural and magnetic phases present in the alloys. Furthermore, INS experiments in MSMAs with different compositions yield specific phonon dispersion curves for each alloy, being therefore these curves specific fingerprints of the alloy being investigated.

4. Summary

The use of neutron-based techniques to complement basic characterization methods commonly employed in the field of MSMAs is of great interest to the scientific community working in this class of materials. The well-established strong interdependence between the main properties of the alloys (martensitic transformation temperatures, maximum deformation, magnetic transition temperatures, etc.) and their crystal lattices and atomic site occupancies makes the use of neutron diffraction a crucial technique to study MSMAs. A huge number of powder and single crystal neutron diffraction experiments have been performed by scientists from everywhere around the globe, yielding with outstanding precision not only the crystalline phases of the alloys, together with their modulations and phase coexistences when relevant, but also the precise atomic site occupancies and magnetic site densities of alloys made of up to six elements. Doping stoichiometric Ni_2MnGa with different elements, the effect of different alloy casting techniques with similar compositions on the alloy performances, or the relevance of Mn-Mn interactions within the magnetic behavior of the alloys are a few examples where neutron diffraction techniques excel.

Small angle neutron scattering and neutron reflectometry experiments, although less numerous than the aforementioned diffraction ones, have also been conducted in MSMAs of different nature. Small angle neutron scattering has been mainly employed to determine phase segregations, both structural and magnetic, as well as to study the presence of certain element-specific rich regions within the MSMAs. Neutron reflectometry has been utilized to investigate the magnetic depth profile of MSMA thin films, being this technique of great potential interest to investigate the interface quality and associated residual stresses in coupled MSMA thin films.

Neutron imaging, and in particular texture diffractometry and radiography (2D) and tomography (3D), provide information about the presence of textures and their preferred orientations, as well as about the in-plane and 3D morphology of twinned regions within the martensitic phases of MSMAs.

In addition to the elastic neutron scattering techniques summarized above, inelastic neutron scattering, namely diffuse scattering experiments, have been performed in MSMAs. These experiments yield valuable information about the phonon dispersion curves for the studied MSMAs, with particular attention to the changes in dispersion events taking place across the phase transformations. All the neutron scattering experiments on MSMAs discussed in the present manuscript have been listed in Table A1, which is presented in Appendix A.

Neutron sources worldwide are continuously being employed to investigate MSMAs, including ILL (Grenoble, France), ISIS (Oxfordshire, UK), FRM-2 (Munich, Germany), NIST (Maryland, USA), SNS (Oak Ridge, USA) and ANSTO (Sydney, Australia).

The use of neutron scattering techniques to study MSMAs suppose great advances in the basic and applied knowledge of the behavior and properties of these interesting alloys. With a future in which the next-generation neutron sources, with improved brilliance, flux and polarization capabilities of their neutron beams, will complement the already existing ones and increase the availability of neutron beams, the interest and relevance of the neutron studies will continue its rise and popularity around the MSMA community.

Author Contributions: Conceptualization, J.M.P.; methodology, N.A.R.-L. and P.L.; writing—original draft preparation, N.A.R.-L. and J.M.P.; writing—review and editing, all authors; supervision, V.C. and J.M.P. All authors have read and agreed to the published version of the manuscript.

Funding: This work has been carried out with the financial support of the Spanish Ministry of Science, Innovation and Universities (project RTI2018-094683-B-C53-54) and Basque Government Department of Education (project IT1245-19). N.A.R.-L. wants to thank the Basque Government (Department of Education) for providing funding under the specific investigation PFPI grant.

Conflicts of Interest: The authors declare no conflict of interest.

Appendix A

Table listing the literature discussed throughout the manuscript, including references, compositions of the alloys studied, first authors, neutron scattering technique employed and an express summary of the investigation carried out in each publication.

Table A1. List of the literature discussed throughout the present review manuscript. PND: powder neutron diffraction; SCND: single crystal neutron diffraction; SANS: small angle neutron scattering; NR: neutron reflectometry; INS: inelastic neutron scattering.

Ref	Composition	First Author	Technique	Express Summary
[18]	NiMnGaCoCuFe	Pérez-Checa	PND	Effect of Fe doping on atomic ordering and magnetic properties
[39]	Ni2MnGa	Brown	PND	MT phase diagram determination
[40]	NiMnGa	Cong	Polarized PND	Differences in the MT in stoichiometric and non-stoichiometric NiMnGa
[41]	NiMnGaCo	Cong	PND, Imaging	Effect of Co doping on crystal structure
[42]	NiMnGaCo	Orlandi	PND	Effect of Co doping on MT
[43]	NiMnGaCo	Orlandi	PND	Effect of Co doping on MT
[44]	NiMnGa	Richard	PND, SCND	Ni and Mn content effect on atomic ordering
[45]	NiMnGa	Lázpita	PND, SCND	Ni and Mn content effect on atomic ordering
[46]	NiMnGa	Lázpita	PND	Ni and Mn content effect on atomic ordering
[47]	NiMnGa	Wang	PND	Study of amorphous phases and MT
[48]	NiMnGaCu	Roy	PND	Effect of Cu doping on TC and TM
[49]	NiMnSn	Brown	PND	Study of atomic ordering and crystal unit cell
[50]	NiMnIn	Mañosa	PND	Study of the magnetic field-induced MT
[51]	Mn2NiGa	Brown	PND	Study of atomic ordering and crystal unit cell
[52]	NiMnSn	Mukadam	PND	Ni and Mn content effect on magnetism
[59]	NiMnGaCu	Glavatskyy	SCND	Cu doping effect on crystal structures, MT and magnetoplasticity
[60]	NiFeGa	Brown	SCND	Effect of applied stress on TC, TM and MT mechanism
[61]	NiMnGa	Glavatskyy	SCND	Study of the MT behavior
[62]	NiMnGa	Molnar	SCND	Study of the stress-induced martensitic variants reorientation
[63]	NiMnGa	Chmielus	SCND	Study of lattice parameters and modulations
[64]	NiMnGa	Chmielus	SCND	Study of lattice parameters and modulations
[65]	NiMnGa	Chmielus	SCND	Study of lattice parameters and modulations
[66]	Ni2MnGa	Kabra	SCND, Imaging	Study of the crystallographic orientation relationship in twinned/untwinned regions
[38]	Ni2MnGa	Webster	Polarized PND	Magnetic site densities in Ni and Mn sites
[68]	Ni2MnGa	Brown	Polarized SCND	Ni to Mn magnetic moment transfer in MT
[69]	NiMnGa	Pramanick	Polarized SCND	Correlation between twin-reorientation and rotation of magnetic moments

Table A1. Cont.

Ref	Composition	First Author	Technique	Express Summary
[70]	NiMnGa	Lázpita	Polarized SCND	Influence of atomic ordering on magnetic coupling between Mn atoms
[74]	NiMnGa	Runov	SANS	Evidence for an intramartensitic phase transition
[75]	NiMnGa	Runov	SANS	Study of spin-spin correlation radius upon cooling to TM
[76]	NiMnCuAl	Sun	SANS	Effect of ageing treatments on Cu- and Mn-rich cluster formation
[77]	NiMnCuAl	Sun	SANS	Effect of ageing treatments on Cu- and Mn-rich cluster formation
[78]	NiMnIn	Benacchio	SANS	Presence of ferromagnetic nanoprecipitates in an antiferromagnetic background
[79]	NiMnCoSn	Sarkar	SANS	Presence of spin-clusters related to MT
[80]	FeMn, FeMnSi, FeMnSiCr, FeMnSiCrNi	Kopitsa	SANS	Effect of Si, Cr and Ni doping and interstitial C and N atoms on MSMA homogeneity
[81]	FeMn, FeMnSi, FeMnSiCr, FeMnSiCrNi	Bliznuk	SANS	Effect of Si, Cr and Ni doping and interstitial C and N atoms on MSMA homogeneity
[87]	NiMnIn	Granovsky	NR	Analysis of induced magnetic moment collinear with the applied field at low T
[92]	NiMnGa	Nie	Imaging	Effect of applied stress on twin variants
[93]	NiMnGa	Chulist	Imaging	Alloy casting method effect on crystal textures
[96]	CoNiGa	Samothrakitis	Imaging	3D microstructure and crystal orientation
[100]	Ni2MnGa	Zheludev	INS	Determination of phonon dispersion curves
[101]	Ni2MnGa	Zheludev	INS	Determination of phonon dispersion curves
[102]	Ni2MnGa	Recarte	INS	Influence of magnetic field on phonon dispersion curves
[103]	NiMnIn	Moya	INS	Determination of phonon dispersion curves
[104]	NiMnX (X = Ga,In,Sn,Sb,Al)	Moya	INS	Determination of phonon dispersion curves

References

1. Heczko, O.; Scheerbaum, N.; Gutfleisch, O. Magnetic shape memory phenomena. *Nanoscale Magn. Mater. Appl.* **2009**, 399–439. [CrossRef]
2. Acet, M.; Mañosa, L.; Planes, A. Magnetic-Field-Induced Effects in Martensitic Heusler-Based Magnetic Shape Memory Alloys. *Handb. Magn. Mater.* **2011**, *19*, 231–289.
3. L'vov, V.A.; Chernenko, V.A.; Barandiaran, J.M. Magnetic shape memory materials with improved functional properties: Scientific aspects. In *Novel Functional Magnetic Materials*; Springer International Publishing: Berlin/Heidelberg, Germany, 2016; Volume 231, pp. 1–40.
4. Chernenko, V. Magnetostrictive Ni-Mn-Based Heusler Alloys. *Ref. Modul. Mater. Sci. Mater. Eng.* **2021**. [CrossRef]
5. Karaca, H.E.; Karaman, I.; Basaran, B.; Ren, Y.; Chumlyakov, Y.I.; Maier, H. J Magnetic field-induced phase transformation in NiMnCoIn magnetic shape-memory alloys-a new actuation mechanism with large work output. *Adv. Funct. Mater.* **2009**, *19*, 983–998. [CrossRef]
6. Faran, E.; Shilo, D. Ferromagnetic Shape Memory Alloys—Challenges, Applications, and Experimental Characterization. *Exp. Tech.* **2016**, *40*, 1005–1031. [CrossRef]
7. Billinge, S.J.L.; Levin, I. The problem with determining atomic structure at the nanoscale. *Science* **2007**, *316*, 561–565. [CrossRef]
8. Ping Liu, J.; Gutfleisch, O.; Fullerton, E.; Sellmyer, D.J. Nanoscale magnetic materials and applications. *Nanoscale Magn. Mater. Appl.* **2009**. [CrossRef]

9. Isnard, O. A review of in situ and/or time resolved neutron scattering. *Comptes Rendus Phys.* **2007**, *8*, 789–805. [CrossRef]
10. Otsuka, K.; Wayman, C.M. *Shape Memory Materials. Unitex*; Cambridge University Press: Cambridge, UK, 1999.
11. Chowdhury, P.; Sehitoglu, H. Deformation physics of shape memory alloys—Fundamentals at atomistic frontier. *Prog. Mater. Sci.* **2017**, *88*, 49–88. [CrossRef]
12. Petrini, L.; Migliavacca, F. Biomedical Applications of Shape Memory Alloys. *J. Metall.* **2011**, *2011*, 1–15. [CrossRef]
13. Kohl, M. *Shape Memory Microactuators*; Springer Science & Business Media: Berlin/Heidelberg, Germany, 2004.
14. Murray, S.J.; Marioni, M.; Allen, S.M.; O'Handley, R.C.; Lograsso, T.A. 6% magnetic-field-induced strain by twin-boundary motion in ferromagnetic Ni-Mn-Ga. *Appl. Phys. Lett.* **2000**, *77*, 886–888. [CrossRef]
15. Chernenko, V.A.; L'vov, V.A.; Cesari, E.; Barandiaran, J.M. Fundamentals of magnetocaloric effect in magnetic shape memory alloys. In *Handbook of Magnetic Materials*; Elsevier: Amsterdam, The Netherlands, 2019; Volume 28, pp. 1–45.
16. Straka, L.; Heczko, O.; Seiner, H.; Lanska, N.; Drahokoupil, J.; Soroka, A.; Fähler, S.; Hänninen, H.; Sozinov, A. Highly mobile twinned interface in 10 M modulated Ni-Mn-Ga martensite: Analysis beyond the tetragonal approximation of lattice. *Acta Mater.* **2011**, *59*, 7450–7463. [CrossRef]
17. Checa, A.P. Development of New Ni–Mn–Ga based High Temperature Shape Memory Alloys. Ph.D. Thesis, University of the Basque Country, Biscay, Spain, 2019.
18. Pérez-Checa, A.; Porro, J.M.; Feuchtwanger, J.; Lázpita, P.; Hansen, T.C.; Mondelli, C.; Sozinov, A.; Barandiarán, J.M.; Ullakko, K.; Chernenko, V. Role of Fe addition in Ni–Mn–Ga–Co–Cu–Fe ferromagnetic shape memory alloys for high-temperature magnetic actuation. *Acta Mater.* **2020**, *196*, 549–555. [CrossRef]
19. Ito, W.; Imano, Y.; Kainuma, R.; Sutou, Y.; Oikawa, K.; Ishida, K. Martensitic and magnetic transformation behaviors in Heusler-type NiMnIn and NiCoMnIn metamagnetic shape memory alloys. *Metall. Mater. Trans. A Phys. Metall. Mater. Sci.* **2007**, *38*, 759–766. [CrossRef]
20. Umetsu, R.Y.; Xu, X.; Kainuma, R. NiMn-based metamagnetic shape memory alloys. *Scr. Mater.* **2016**, *116*, 1–6. [CrossRef]
21. Kainuma, R.; Imano, Y.; Ito, W.; Sutou, Y.; Morito, H.; Okamoto, S.; Kitakami, O.; Oikawa, K.; Fujita, A.; Kanomata, T.; et al. Magnetic-field-induced shape recovery by reverse phase transformation. *Nature* **2006**, *439*, 957–960. [CrossRef] [PubMed]
22. Krenke, T.; Duman, E.; Acet, M.; Wassermann, E.F.; Moya, X.; Manosa, L.; Planes, A. Inverse magnetocaloric effect in ferromagnetic Ni-Mn-Sn alloys. *Nat. Mater.* **2005**, *4*, 450–454. [CrossRef]
23. Dunand, D.C.; Müllner, P. Size effects on magnetic actuation in Ni-Mn-Ga shape-memory alloys. *Adv. Mater.* **2011**, *23*, 216–232. [CrossRef]
24. Wang, J.; Jiang, C.; Techapiesancharoenkij, R.; Bono, D.; Allen, S.M.; O'Handley, R.C. Microstructure and magnetic properties of melt spinning Ni-Mn-Ga. *Intermetallics* **2013**, *32*, 151–155. [CrossRef]
25. Jones, H. Chapter 3 Rapid solidification. *Pergamon Mater. Ser.* **1999**, *2*, 23–45.
26. Chernenko, V.A.; Kokorin, V.V.; Vitenko, I.N. Properties of ribbon made from shape memory alloy Ni2MnGa by quenching from the liquid state. *Smart Mater. Struct.* **1994**, *3*, 80–82. [CrossRef]
27. Gaitzsch, U.; Pötschke, M.; Roth, S.; Rellinghaus, B.; Schultz, L. A 1% magnetostrain in polycrystalline 5M Ni-Mn-Ga. *Acta Mater.* **2009**, *57*, 365–370. [CrossRef]
28. Aseguinolaza, I.R.; Orue, I.; Svalov, A.V.; Wilson, K.; Müllner, P.; Barandiarán, J.M.; Chernenko, V.A. Martensitic transformation in Ni-Mn-Ga/Si(100) thin films. *Thin Solid Film* **2014**, *558*, 449–454. [CrossRef]
29. Lambrecht, F.; Lay, C.; Aseguinolaza, I.R.; Chernenko, V.; Kohl, M. NiMnGa/Si Shape Memory Bimorph Nanoactuation. *Shape Mem. Superelasticity* **2016**, *2*, 347–359. [CrossRef]
30. Bailey, I.F. A review of sample environments in neutron scattering. *Z. Fur Krist.* **2003**, *218*, 84–95. [CrossRef]
31. Sears, V.F. Neutron scattering lengths and cross sections. *Neutron News* **1992**, *3*, 26–37. [CrossRef]
32. Pynn, R. *Neutron Scattering Primer*; Los Alamos Science: Los Alamos, NM, USA, 1990; Volume 19.
33. Willis, B.T.M.; Carlile, C.J. Experimental Neutron Scattering. *Anim. Genet.* **2009**, *39*, 561–563.
34. Orench, I.P.; Clergeau, J.F.; Martínez, S.; Olmos, M.; Fabelo, O.; Campo, J. The new powder diffractometer D1B of the Institut Laue Langevin. *J. Phys. Conf. Ser.* **2014**, *549*, 12003. [CrossRef]
35. Aubert, A.; Puente-Orench, I.; Porro, J.M.; Luca, S.; Garitaonandia, J.S.; Barandiaran, J.M.; Hadjipanayis, G.C. Denitrogenation process in ThMn$_{12}$ nitride by *in situ* neutron powder diffraction. *Phys. Rev. Mat.* **2021**, *5*, 014415.
36. Baruchel, J.; Hodeau, J.L.; Lehmann, M.S.; Regnard, J.R.; Schlenker, C. *Neutron and Synchrotron Radiation for Condensed Matter Studies. Neutron and Synchrotron Radiation for Condensed Matter Studies*; EDP Sciences; Springer: Berlin/Heidelberg, Germany, 1994. [CrossRef]
37. Albinati, A.; Willis, B.T.M. The Rietveld method in neutron and X-ray powder diffraction. *J. Appl. Crystallogr.* **1982**, *15*, 361–374. [CrossRef]
38. Webster, P.J.; Ziebeck, K.R.A.; Town, S.L.; Peak, M.S. Magnetic order and phase transformation in Ni2MnGa. *Philos. Mag. B* **1984**, *49*, 295–310. [CrossRef]
39. Brown, P.J.; Crangle, J.; Kanomata, T.; Matsumoto, M.; Neumann, K.-U.; Ouladdiaf, B.; Ziebeck, K. The crystal structure and phase transitions of the magnetic shape memory compound Ni 2 MnGa. *J. Phys. Condens. Matter.* **2002**, *14*, 10159–10171. [CrossRef]
40. Cong, D.Y.; Zetterström, P.; Wang, Y.D.; Delaplane, R.; Peng, R.L.; Zhao, X.; Zuo, L. Crystal structure and phase transformation in Ni 53Mn 25Ga 22 shape memory alloy from 20 K to 473 K. *Appl. Phys. Lett.* **2005**, *87*, 85–88. [CrossRef]

41. Cong, D.Y.; Wang, Y.D.; Lin Peng, R.; Zetterström, P.; Zhao, X.; Liaw, P.K.; Zuo, L. Crystal structures and textures in the hot-forged Ni-Mn-Ga shape memory alloys. *Metall. Mater. Trans. A Phys. Metall. Mater. Sci.* **2006**, *37*, 1397–1403. [CrossRef]
42. Orlandi, F.; Çaklr, A.; Manuel, P.; Khalyavin, D.D.; Acet, M.; Righi, L. Neutron diffraction and symmetry analysis of the martensitic transformation in Co-doped Ni2MnGa. *Phys. Rev. B* **2020**, *101*, 1–13. [CrossRef]
43. Orlandi, F.; Fabbrici, S.; Albertini, F.; Manuel, P.; Khalyavin, D.D.; Righi, L. Long-range antiferromagnetic interactions in Ni-Co-Mn-Ga metamagnetic Heusler alloys: A two-step ordering studied by neutron diffraction. *Phys. Rev. B* **2016**, *94*, 1–5. [CrossRef]
44. Richard, M.L.; Feuchtwanger, J.; Allen, S.M.; O'Handley, R.C.; Lázpita, P.; Barandiaran, J.M.; Gutierrez, J.; Ouladdiaf, B.; Mondelli, C.; Lograsso, T.; et al. Chemical order in off-stoichiometric Ni-Mn-Ga ferromagnetic shape-memory alloys studied with neutron diffraction. *Philos. Mag.* **2007**, *87*, 3437–3447. [CrossRef]
45. Lázpita, P.; Barandiarán, J.M.; Gutiérrez, J.; Richard, M.; Allen, S.M.; O'Handley, R.C. Magnetic and structural properties of non-stoichiometric Ni-Mn-Ga ferromagnetic shape memory alloys. *Eur. Phys. J. Spec. Top.* **2008**, *158*, 149–154. [CrossRef]
46. Lázpita, P.; Barandiarán, J.M.; Gutiérrez, J.; Feuchtwanger, J.; Chernenko, V.A.; Richard, M.L. Magnetic moment and chemical order in off-stoichiometric Ni–Mn–Ga ferromagnetic shape memory alloys. *New J. Phys.* **2011**, *13*, 033039. [CrossRef]
47. Wang, Y.D.; Ren, Y.; Nie, Z.H.; Liu, D.M.; Zuo, L.; Choo, H.; Li, H.; Liaw, P.K.; Yan, J.Q.; McQueeney, R.J.; et al. Structural transition of ferromagnetic Ni2MnGa nanoparticles. *J. Appl. Phys.* **2007**, *101*, 1–7. [CrossRef]
48. Roy, S.; Blackburn, E.; Valvidares, S.M.; Fitzsimmons, M.R.; Vogel, S.C.; Khan, M.; Dubenko, I.; Stadler, S.; Ali, N.; Sinha, S.K.; et al. Delocalization and hybridization enhance the magnetocaloric effect in Cu-doped Ni2 MnGa. *Phys. Rev. B Condens. Matter. Mater. Phys.* **2009**, *79*, 1–5. [CrossRef]
49. Brown, P.J.; Gandy, A.P.; Ishida, K.; Kainuma, R.; Kanomata, T.; Neumann, K.U.; Oikawa, K.; Ouladdiaf, B.; Ziebeck, K.R.A. The magnetic and structural properties of the magnetic shape memory compound Ni2Mn1.44Sn0.56. *J. Phys. Condens. Matter.* **2006**, *18*, 2249–2259. [CrossRef]
50. Mañosa, L.; Moya, X.; Planes, A.; Krenke, T.; Acet, M.; Wassermann, E.F. Ni-Mn-based magnetic shape memory alloys: Magnetic properties and martensitic transition. *Mater. Sci. Eng. A* **2008**, *481-482*, 49–56. [CrossRef]
51. Brown, P.J.; Kanomata, T.; Neumann, K.; Neumann, K.U.; Ouladdiaf, B.; Sheikh, A.; Ziebeck, K.R.A. Atomic and magnetic order in the shape memory alloy Mn2NiGa. *J. Phys. Condens. Matter* **2010**, *22*, 506001. [CrossRef] [PubMed]
52. Mukadam, M.D.; Yusuf, S.M.; Bhatt, P. Tuning the magnetocaloric properties of the Ni 2 × Mn 1 − X Sn Heusler alloys. *J. Appl. Phys.* **2013**, *113*, 1–6. [CrossRef]
53. Lelièvre-Berna, E.; Bourgeat-Lami, E.; Gibert, Y.; Kernavanois, N.; Locatelli, J.; Mary, T.; Pastrello, G.; Petukhov, A.; Pujol, S.; Rouques, R.; et al. ILL polarised hot-neutron beam facility D3. In *Physica B: Condensed Matter*; Elsevier: North-Holland, The Netherlands, 2005; Volume 356, pp. 141–145.
54. Goldman, A.I. Neutron Techniques. In *Characterization of Materials*; Kaufmann, E.N., Ed.; John Wiley & Sons: Hoboken, NJ, USA, 2002; pp. 2192–2204. [CrossRef]
55. Chmielus, M. Composition, Structure and Magneto-Mechanical Properties of Ni-Mn-Ga Magnetic Shape-Memory Alloys. Ph.D. Thesis, Technische Universität Berlin, Berlin, Germany, 2010. [CrossRef]
56. Long, G.J. Neutron Diffraction. In *Comprehensive Coordination Chemistry II*; Clarendon Press: Wotton-under-Edge, UK, 2004; Volume 2.
57. Artioli, G. Single-crystal neutron diffraction. *Eur. J. Mineral* **2002**, *14*, 233–239. [CrossRef]
58. Rodríguez-Carvajal, J. Recent advances in magnetic structure determination by neutron powder diffraction. *Phys. B Condens. Matter* **1993**, *192*, 55–69. [CrossRef]
59. Glavatskyy, I.; Glavatska, N.; Dobrinsky, A.; Hoffmann, J.U.; Söderberg, O.; Hannula, S.P. Crystal structure and high-temperature magnetoplasticity in the new Ni-Mn-Ga-Cu magnetic shape memory alloys. *Scr. Mater.* **2007**, *56*, 565–568. [CrossRef]
60. Brown, P.J.; Gandy, A.P.; Ishida, K.; Kainuma, R.; Kanomata, T.; Morito, H.; Neumann, K.U.; Oikawa, K.; Ziebeck, K.R.A. Crystal structures and magnetization distributions in the field dependent ferromagnetic shape memory alloy Ni54Fe19Ga27. *J. Phys. Condens. Matter* **2007**, *19*, 016201. [CrossRef]
61. Glavatskyy, I.; Glavatska, N.; Urubkov, I.; Hoffman, J.U.; Bourdarot, F. Crystal and magnetic structure temperature evolution in Ni-Mn-Ga magnetic shape memory martensite. *Mater. Sci. Eng. A* **2008**, *481-482*, 298–301. [CrossRef]
62. Molnar, P.; Sittner, P.; Lukas, P.; Hannula, S.P.; Heczko, O. Stress-induced martensite variant reorientation in magnetic shape memory Ni-Mn-Ga single crystal studied by neutron diffraction. *Smart Mater. Struct.* **2008**, *17*, 035014. [CrossRef]
63. Chmielus, M.; Glavatskyy, I.; Hoffmann, J.U.; Chernenko, V.A.; Schneider, R.; Müllner, P. Influence of constraints and twinning stress on magnetic field-induced strain of magnetic shape-memory alloys. *Scr. Mater.* **2011**, *64*, 888–891. [CrossRef]
64. Chmielus, M.; Rolfs, K.; Wimpory, R.; Reimers, W.; Müllner, P.; Schneider, R. Effects of surface roughness and training on the twinning stress of Ni-Mn-Ga single crystals. *Acta Mater.* **2010**, *58*, 3952–3962. [CrossRef]
65. Chmielus, M.; Witherspoon, C.; Wimpory, R.C.; Paulke, A.; Hilger, A.; Zhang, X.; Dunand, D.C.; Müllner, P. Magnetic-field-induced recovery strain in polycrystalline Ni-Mn-Ga foam. *J. Appl. Phys.* **2010**, *108*. [CrossRef]
66. Kabra, S.; Kelleher, J.; Kockelmann, W.; Gutmann, M.; Tremsin, A. Energy-dispersive neutron imaging and diffraction of magnetically driven twins in a Ni2MnGa single crystal magnetic shape memory alloy. *J. Phys. Conf. Ser.* **2016**, *746*, 012056. [CrossRef]
67. Hicks, T.J. Experiments with neutron polarization analysis. *Adv. Phys.* **1996**, *45*, 243–298. [CrossRef]

68. Brown, P.J.; Bargawi, A.Y.; Crangle, J.; Neumann, K.U.; Ziebeck, K.R.A. Direct observation of a band Jahn-Teller effect in the martensitic phase transition of Ni2MnGa. *J. Phys. Condens. Matter* **1999**, *11*, 4715–4722. [CrossRef]
69. Pramanick, A.; Glavic, A.; Samolyuk, G.; Aczel, A.A.; Lauter, V.; Ambaye, H.; Gai, Z.; Ma, J.; Stoica, A.D.; Stocks, G.M.; et al. Direct in situ measurement of coupled magnetostructural evolution in a ferromagnetic shape memory alloy and its theoretical modeling. *Phys. Rev. B Condens. Matter Mater. Phys.* **2015**, *92*, 1–12. [CrossRef]
70. Lázpita, P.; Barandiarán, J.M.; Gutiérrez, J.; Mondelli, C.; Sozinov, A.; Chernenko, V.A. Polarized Neutron Study of Ni-Mn-Ga Alloys: Site-Specific Spin Density Affected by Martensitic Transformation. *Phys. Rev. Lett.* **2017**, *119*, 155701. [CrossRef]
71. Heenan, R.K.; Rogers, S.E.; Turner, D.; Terry, A.E.; Treadgold, J.; King, S.M. Small angle neutron scattering using sans2d. *Neutron News* **2011**, *22*, 19–21. [CrossRef]
72. Boothroyd, A.T. The effect of gravity on the resolution of small-angle neutron scattering. *J. Appl. Crystallogr.* **1989**, *22*, 252–255. [CrossRef]
73. Antony, A.; Schmerl, N.M.; Sokolova, A.; Mahjoub, R.; Fabijanic, D.; Stanford, N.E. Quantification of the dislocation density, size, and volume fraction of precipitates in deep cryogenically treated martensitic steels. *Metals* **2020**, *10*, 1561. [CrossRef]
74. Runov, V.V.; Chernenkov, Y.P.; Runova, M.K.; Gavriljuk, V.G.; Glavatska, N.I. Study of phase transitions and mesoscopic magnetic structure in Ni-Mn-Ga by means of small-angle polarized neutron scattering. *Phys. B Condens. Matter* **2003**, *335*, 109–113. [CrossRef]
75. Runov, V.; Runova, M.; Gavriljuk, V.; Glavatska, N. Observation of magnetic-nuclear cross-correlations in Ni-Mn-Ga. *Phys. B Condens. Matter* **2004**, *350*, 87–89. [CrossRef]
76. Sun, L.Y.; Vasin, R.N.; Islamov, A.K.; Bobrikov, I.A.; Cifre, J.; Golovin, I.S.; Balagurov, A.M. Influence of spinodal decomposition on structure and thermoelastic martensitic transition in MnCuAlNi alloy. *Mater. Lett.* **2020**, *275*, 128069. [CrossRef]
77. Sun, L.; Sumnikov, S.V.; Islamov, A.K.; Vasin, R.N.; Bobrikov, I.A.; Balagurov, A.M.; Cheng, W.; Churyumov, Y.; Golovin, I.S. Spinodal decomposition influence of austenite on martensitic transition in a Mn-13 at.%Cu alloy. *J. Alloys Compd.* **2021**, *853*, 157061. [CrossRef]
78. Benacchio, G.; Titov, I.; Malyeyev, A.; Peral, I.; Bersweiler, M.; Bender, P.; Mettus, D.; Honecker, D.; Gilbert, E.P.; Coduri, M.; et al. Evidence for the formation of nanoprecipitates with magnetically disordered regions in bulk Ni50Mn45In5 Heusler alloys. *Phys. Rev. B* **2019**, *99*, 184422. [CrossRef]
79. Sarkar, S.K.; Ahlawat, S.; Kaushik, S.D.; Babu, P.D.; Sen, D.; Honecker, D.; Biswas, A. Magnetic ordering of the martensite phase in Ni-Co-Mn-Sn-based ferromagnetic shape memory alloys. *arXiv* **2019**, arXiv:1908.08860v2. [CrossRef]
80. Kopitsa, G.P.; Runov, V.V.; Grigoriev, S.V.; Bliznuk, V.V.; Gavriljuk, V.G.; Glavatska, N.I. The investigation of Fe-Mn-based alloys with shape memory effect by small-angle scattering of polarized neutrons. *Phys. B Condens. Matter* **2003**, *335*, 134–139. [CrossRef]
81. Bliznuk, V.V.; Gavriljuk, V.G.; Kopitsa, G.P.; Grigoriev, S.V.; Runov, V.V. Fluctuations of chemical composition of austenite and their consequence on shape memory effect in Fe-Mn-(Si, Cr, Ni, C, N) alloys. *Acta Mater.* **2004**, *52*, 4791–4799. [CrossRef]
82. Webster, J.R.P.; Langridge, S.; Dalgliesh, R.M.; Charlton, T.R. Reflectometry techniques on the second target station at ISIS: Methods and science. *Eur. Phys. J. Plus* **2011**, *126*, 1–5. [CrossRef]
83. Daillant, J.; Gibaud, A. *X-Ray and Neutron Reflectivity: Principles and Applications*; Springer: Berlin/Heidelberg, Germany, 1999.
84. Tolan, M.; Press, W. X-ray and neutron reflectivity. *Z. Fur Krist.* **1998**, *213*, 319–336. [CrossRef]
85. Khaydukov, Y.N.; Kravtsov, E.A.; Zhaketov, V.D.; Progliado, V.V.; Kim, G.; Nikitenko, Y.V.; Keller, T.; Ustinov, V.V.; Aksenov, V.L.; Keimer, B. Magnetic proximity effect in Nb/Gd superlattices seen by neutron reflectometry. *Phys. Rev. B* **2019**, *99*, 3–7. [CrossRef]
86. Singh, S.; Swain, M.; Basu, S. Kinetics of interface alloy phase formation at nanometer length scale in ultra-thin films: X-ray and polarized neutron reflectometry. *Prog. Mater. Sci.* **2018**, *96*, 1–50. [CrossRef]
87. Granovsky, S.; Gaidukova, I.; Sokolov, A.; Devishvili, A.; Snegirev, V. Structural and magnetic properties of Ni50Mn35In15 thin films. *Solid State Phenom.* **2015**, *233–234*, 666–669. [CrossRef]
88. Hussey, D.S.; Brocker, C.; Cook, J.C.; Jacobson, D.L.; Gentile, T.R.; Chen, W.C.; Baltic, E.; Baxter, D.V.; Doskow, J.; Arif, M. A New Cold Neutron Imaging Instrument at NIST. In *Physics Procedia*; Elsevier B.V.: Amsterdam, The Netherlands, 2015; Volume 69, pp. 48–54.
89. Wenk, H.R.; Kern, H.; Schaefer, W.; Will, G. Comparison of neutron and X-ray diffraction in texture analysis of deformed carbonate rocks. *J. Struct. Geol.* **1984**, *6*, 687–692. [CrossRef]
90. Matthies, S.; Pehl, J.; Wenk, H.R.; Lutterotti, L.; Vogel, S.C. Quantitative texture analysis with the HIPPO neutron TOF diffractometer. *J. Appl. Crystallogr.* **2005**, *38*, 462–475. [CrossRef]
91. Brokmeier, H.G.; Gan, W.M.; Randau, C.; Völler, M.; Rebelo-Kornmeier, J.; Hofmann, M. Texture analysis at neutron diffractometer STRESS-SPEC. *Nucl. Instrum. Methods Phys. Res. Sect. A Accel. Spectrometers Detect. Assoc. Equip.* **2011**, *642*, 87–92. [CrossRef]
92. Nie, Z.H.; Wang, Y.D.; Wang, G.Y.; Richardson, J.W.; Wang, G.; Liu, Y.D.; Liaw, P.K.; Zuo, L. Phase transition and texture evolution in the Ni-Mn-Ga ferromagnetic shape-memory alloys studied by a neutron diffraction technique. *Metall. Mater. Trans. A Phys. Metall. Mater. Sci.* **2008**, *39*, 3113–3119. [CrossRef]
93. Chulist, R.; Poetschke, M.; Boehm, A.; Brokmeier, H.G.; Garbe, U.; Lippmann, T.; Oertel, C.G.; Skrotzki, W. Cast and rolling textures of NiMnGa alloys. *Mater. Res. Soc. Symp. Proc.* **2008**, *1050*, 30–35. [CrossRef]
94. Winkler, B. Applications of neutron radiography and neutron tomography. *Rev. Mineral. Geochem.* **2006**, *63*, 459–471. [CrossRef]
95. Woracek, R.; Penumadu, D.; Kardjilov, N.; Hilger, A.; Boin, M.; Banhart, J.; Manke, I. 3D mapping of crystallographic phase distribution using energy-selective neutron tomography. *Adv. Mater.* **2014**, *26*, 4069–4073. [CrossRef] [PubMed]

96. Samothrakitis, S.; Larsen, C.B.; Woracek, R.; Heller, L.; Kopeček, J.; Gerstein, G.; Maier, H.J.; Rameš, M.; Tovar, M.; Šittner, P.; et al. A multiscale study of hot-extruded CoNiGa ferromagnetic shape-memory alloys. *Mater. Des.* **2020**, *196*, 109118. [CrossRef]
97. Jiménez-Ruiz, M.; Ivanov, A.; Fuard, S. LAGRANGE—The new neutron vibrational spectrometer at the ILL. *J. Phys. Conf. Ser.* **2014**, *549*, 012004. [CrossRef]
98. Eckert, J. Theoretical introduction to neutron scattering spectroscopy. *Spectrochim. Acta Part A Mol. Spectrosc.* **1992**, *48*, 271–283. [CrossRef]
99. Hudson, B.S. Inelastic Neutron Scattering: A Tool in Molecular Vibrational Spectroscopy and a Test of ab Initio Methods. *J. Phys. Chem. A* **2001**, *105*, 3949–3960. [CrossRef]
100. Zheludev, A.; Shapiro, S.M.; Wochner, P.; Schwartz, A.; Wall, M.; Tanner, L.E. Phonon anomaly, central peak, and microstructures in Ni2MnGa. *Phys. Rev. B* **1995**, *51*, 11310–11314. [CrossRef] [PubMed]
101. Zheludev, A.; Shapiro, S.; Wochner, P.; Tanner, L. Precursor effects and premartensitic transformation inMnGa. *Phys. Rev. B Condens. Matter Mater. Phys.* **1996**, *54*, 15045–15050. [CrossRef]
102. Recarte, V.; Pérez-Landazábal, J.I.; Sánchez-Alarcos, V.; Cesari, E.; Jiménez-Ruiz, M.; Schmalzl, K.; Chernenko, V.A. Direct evidence of the magnetoelastic interaction in Ni2MnGa magnetic shape memory system. *Appl. Phys. Lett.* **2013**, *102*, 1–5. [CrossRef]
103. Moya, X.; González-Alonso, D.; Mañosa, L.; Planes, A.; Garlea, V.O.; Lograsso, T.A.; Schlagel, D.L.; Zarestky, J.L.; Aksoy, S.; Acet, M. Lattice dynamics in magnetic superelastic Ni-Mn-In alloys: Neutron scattering and ultrasonic experiments. *Phys. Rev. B Condens. Matter Mater. Phys.* **2009**, *79*, 214118. [CrossRef]
104. Moya, X.; Mañosa, L.; Planes, A.; Krenke, T.; Acet, M.; Garlea, V.O.; Lograsso, T.A.; Schlagel, D.L.; Zarestky, J.L. Lattice dynamics and phonon softening in Ni-Mn-Al Heusler alloys. *Phys. Rev. B Condens. Matter Mater. Phys.* **2006**, *73*, 64303. [CrossRef]

Article

Compositional Dependence of Magnetocrystalline Anisotropy in Fe-, Co-, and Cu-Alloyed Ni-Mn-Ga

Michal Rameš, Vít Kopecký and Oleg Heczko *

FZU—Institute of Physics of the Czech Academy of Sciences, Na Slovance 1999/2, 182 21 Prague, Czech Republic; ramesm@fzu.cz (M.R.); kopeckyv@fzu.cz (V.K.)
* Correspondence: heczko@fzu.cz

Abstract: The key for the existence of magnetic induced reorientation is strong magnetocrystalline anisotropy, i.e., the coupling between ferroelastic and ferromagnetic ordering. To increase the transformation temperatures and thus functionality, various elemental alloying in Ni-Mn-Ga is tried. We analyzed more than twenty polycrystalline alloys alloyed by small amount (up to 5atom%) of transitional metals Co, Fe, Ni, and Cu for the value of magnetic anisotropy in search of general trends with alloying. In agreement with previous reports, we found that maximum anisotropy occurs at stoichiometric Ni_2MnGa and any alloying decreases its value. The strongest decrease of the anisotropy is observed in the case where the alloyed elements substitute Ga.

Keywords: magnetocrystalline anisotropy; magnetic shape memory alloys; Ni-Mn-Ga

1. Introduction

Magnetocrystalline anisotropy is a key parameter for existence of magnetically induced reorientation (MIR) [1], i.e., giant magnetic field induced strain enabled by twin boundary motion and consequent lattice reorientation [2]. The magnetocrystalline anisotropy represents a coupling between ferroelastic and ferromagnetic ordering or between magnetization and lattice. If the anisotropy, i.e., the coupling, is weak, the magnetization vector rotates toward the direction of external magnetic field and no reorientation occurs. If, on the other hand, the coupling is strong, crystal lattice follows the magnetization and structural reorientation takes place.

Such structural reorientation was observed already some time ago in Dy and Tb compounds [3] at low temperatures and a very high magnetic field of tens of Tesla. In these compounds, the anisotropy is very large $\approx 10^7$ J/m^3. Apart from the observed peculiar magnetization loops, now recognized as a result of the structural reorientation and the observation of emerging structural twinning, the effect was not fully understood and investigated further. A new impetus was provided by discovery of giant magnetic field induced strain (MFIS) in Ni-Mn-Ga in relatively modest field below 1 T close to room temperature [4].

The importance of magnetocrystalline anisotropy for MIR as one of the magnetic shape memory (MSM) effects [1] was recognized from the onset [5,6] and its magnitude was established for several compositions of Ni-Mn-Ga [7–9]. It was shown that the anisotropy is larger for the stoichiometry alloy and it decreases with increasing Mn content [9]. Although the large MSM effect is possible only in single crystal, the extensive search for improved performance of MSM alloys was done mostly on polycrystalline alloys [10–12]. The main focus is usually on transformation behavior and search for increasing transformation temperatures, i.e., ferromagnetic and martensitic transformation temperatures with various elemental alloying of parent Ni-Mn-Ga compound [13–17]. Such an investigation usually leaves the magnetic anisotropy untouched.

The omission is understandable as the absolute value of the anisotropy can be correctly evaluated only in single crystals [18] and in the case of ferroelastic materials, even the single twin variant single crystal is needed. This was exemplarily demonstrated in single crystal Ni-Mn-Ga NM martensite [19] and in materials alloyed by Cu and Co [20]. The preparation of a single crystal would be quite ineffective for each tried composition, hence the other method is needed in order to provide at least approximate values and to catch the anisotropy trend with various alloying.

Albertini et al. used a method based on singular point detection [21]. However, the magnetic curve measurements may not be always susceptible to this analysis. It can be difficult to locate the expected minimum on second derivative. The anisotropy was also investigated by ferromagnetic resonance (FMR) method [22,23], but again, the polycrystalline and polytwinned martensite can result in a much lower anisotropy value [24] compared to single crystal due to averaging. Another method is to use magnetization approach to saturation [3]. Using this method, the magnetocrystalline anisotropy of polycrystalline Ni_2MnGa was determined to be much higher [25] compared to the anisotropy obtained from the measurement of single crystal [26]. The ab initio computation provides a way to determine magnetocrystalline anisotropy theoretically but their results when compared with the available experiment are not persuasive [20,27]. This is mostly due to fact that a relatively small anisotropy term is determined from the difference of two large numbers, which is prone to significant error.

In summary, all methods can have some merit but in general the value of magnetocrystalline anisotropy determined from polycrystalline materials can exhibit large error and comparison between methods can be questionable. Moreover, other effects contributing to anisotropy measurement (demagnetization, internal stress) are often not considered. Although the absolute value of the anisotropy in polycrystalline materials can be difficult to determine, the comparison of the samples with various alloying but prepared and measured by the same methods can provide good measure of anisotropy trends if not precise quantitative values.

Here we present the evaluation of magnetization energy obtained from magnetization curves of polycrystalline Ni-Mn-Ga alloyed with various content of Fe, Cu, and to smaller extent of Co, covering the area explored in the search to increase the transformation temperatures [28,29]. We assume that the magnetization energy corresponds to effective magnetocrystalline anisotropy. We evaluate the anisotropy energy at 10 K to obtain fundamental value independent from Curie transformation temperature. Although the method does not provide precise values of the anisotropy, it supplies the useful comparative values and helps to evaluate anisotropy trend with elemental alloying. Such knowledge has a practical impact for the MIR effect, in particular alloys and for theoretical investigation, to guide their calculation being able to predict new magnetic shape memory (MSM) materials.

2. Materials and Methods

We prepared two sets of doped materials starting with stoichiometric Ni_2MnGa. The first group contains nine alloys with the substitutions by transitional metals Ni, Co and Fe as in [28]. The second group contains 12 Ni-Mn-Ga-Fe-Cu alloys with varied composition around reference slightly off-stoichiometric ratio $Ni_{50}Mn_{28}Ga_{22}$. The alloys were designed to follow constraints predicted to be important for MSM effect: keeping Ni (Mn) concentration near 50% (28.5%), altering the Fe/Cu ratio. The effort to keep 50% of Ni was not fully successful and some deviations were detected. All alloys were prepared by arc-melting of pure metals under 4×10^{-4} mbar argon atmosphere with Edmund Bühler MAM-1 arc furnace in a water-cooled copper crucible. Ingots were re-melted at least three times to improve homogeneity.

Several samples were annealed in an alumina crucible within a tube-furnace under argon gas flow at 1273 K for 72 h and ordered at 1073 K for 24 h, then left in the furnace to cool slowly. This treatment resulted in 3% Mn loss compared to nominal composition. To keep the composition closer to a nominal one, the majority of samples were annealed in argon-backfilled quartz ampoules. These experienced less than 0.5% Mn loss. Wire electric discharge machining (ZAP BP) was used for sample cutting to obtain samples suitable for magnetic measurement. These samples were cut from the central part of the ingot and have the same shape with dimension approximately 5 × 3 × 1 mm³.

The surfaces of all samples were ground with progressively finer grit SiC papers to remove kerfs formed during discharge cutting and brass contamination from cutting wire. To determine precise elemental compositions, all samples were measured by X-ray fluorescent spectroscopy (XRF) with an Eagle III EDAX μProbe (Roentgenanalytik Systeme GmbH & Co., Taunusstein, Germany). The measurement error of the XRF was highest for Mn and Ga, reaching ±0.5 at. %. The experimentally determined compositions of these alloys are reported in Table 1. We were not able to determine the precise structure of martensite at 10 K due to experimental method limitation. The structure of the samples at 250 K or room temperature are listed in [28,29].

Table 1. Effective magnetic anisotropy K_{eff} at 10 K calculated from magnetization curves in Ni-Mn-Ga alloyed by small amount of Fe and Cu. Experimental composition (at.%) determined by XRF is quoted. All samples listed in the table are in martensite state at 10 K.

No.	Ni (at.%)	Mn (at.%)	Ga (at.%)	Fe (at.%)	Cu (at.%)	K_{eff} at 10 K ($\times 10^5$ J/m³)
1	50.4	24.8	21.7	1.4	1.7	1.2
2	50.0	24.7	20.3	2.4	2.7	1.5
3	49.2	23.7	18.0	4.3	4.7	1.6
4	48.1	27.5	23.6	0.0	0.9	1.9
5	48.1	30.7	17.8	2.7	0.7	1.8
6	47.4	30.9	16.0	4.9	0.8	1.3
7	46.4	24.9	24.6	3.2	0.9	2.1
8	46.5	24.7	24.7	3.2	0.9	2.3
9	46.2	24.9	24.8	3.3	0.9	2.0
10	46.3	26.8	22.8	3.2	0.8	2.1
11	46.2	29.1	20.6	3.3	0.9	1.1
12	46.7	24.5	23.8	3.2	1.8	2.4

Magnetization loops were measured using a vibrating sample magnetometer in PPMS, (Quantum Design, Inc., San Diego, CA, USA). For representation of the ground state of the martensite in our alloys the magnetic hysteresis loops were measured up to 9 T at temperature $T = 10$ K. The measured curves were corrected for demagnetization using slope determined from the curve of cubic austenite measured below T_C, since in (cubic) ferromagnetic austenite the anisotropy is expected to be negligible [7]. The correction is demonstrated in Figure 1. Then descending branches of the corrected curves were numerically integrated with respect to magnetization from remnant magnetization up to saturation magnetization M_s. We put M_s as magnetization at field $\mu_0 H = 3$ T. We marked the calculated magnetic energy as effective anisotropy K_{eff}. Assuming that no other anisotropies are present and after the correction for demagnetization we obtained value comparable to magnetocrystalline anisotropy.

In addition, we try to use singular point detection method for comparison. The minimum in second derivative of the magnetization curve should occur at the anisotropy field. Most of the curves did not, however, exhibit expected minimum, and the anisotropy field could not be determined. In a few available cases, the anisotropy determined by this method was somehow higher than the values determined using magnetic energy. This is to be expected as the singular point detection method should identify the anisotropy field directly.

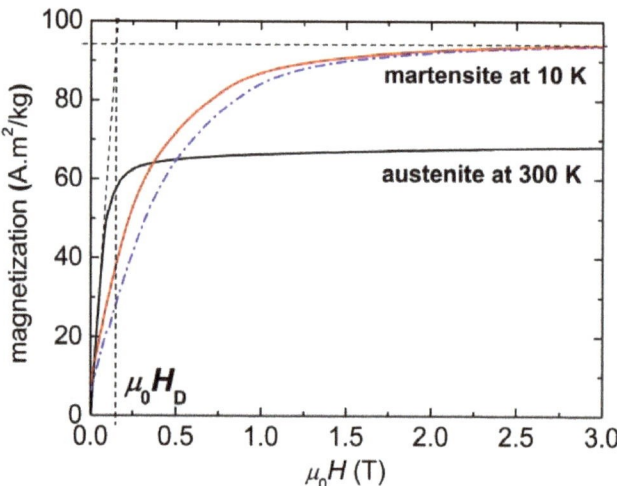

Figure 1. Measured magnetization loops of austenite (black) and martensite (dashed blue) measured at 300 K and 10 K, respectively. Demagnetization field $\mu_0 H_D$ at M_s of martensite determined from the initial slope of the austenite loop is shown. Work done by internal magnetic field, equal to anisotropy energy, is given by the area above the magnetization curve corrected for demagnetization (red).

3. Results and Discussion

Before investigation of the compositional variation of the effective magnetocrystalline anisotropy K_{eff}, we checked the microstructure of the prepared polycrystalline bulk. By arc-melting method, the resulting small ingots were textured with relatively large grains. The grains grow further after annealing. An example of grain orientation determined by EBSD is shown in Figure 2 together with sample cut. The prepared small ingots were heavily textured with [001] preferential columnar orientation from the ingot bottom. The orientation of magnetic field in magnetic measurement was approximately perpendicular to the texture. The samples with large preferentially oriented grains are not very suitable for the singular point detection and approach to saturation methods. Instead, we used evaluation of magnetic energy assuming that the distribution of the grains was similar in all samples.

In addition, after martensitic transformation from cubic to lower symmetry phase the crystal became twinned. Assuming the simplest (pseudo)tetragonal structure, there are three structural variants with c-axis approximately along axes of cubic parent phase connected by twinning. To keep structural compatibility and the same shape, the distribution of twin domains is expected to be about equal, i.e., each orientation of structural or ferroelastic variant occupies one third of the volume. Thus, we can expect that maximum one third of the volume has easy magnetization axis along the field but as the orientation in perpendicular texture direction is random, the probability of easy axis along field is even less. The unknown orientation of easy axes in individual grains is the main source of the error in determining the anisotropy from the curve. It is, however, clear from the sample textured state and nature of the measurement that the determined value of the anisotropy must be smaller than value of a single crystal. Importantly, the Ni-Mn-Ga martensites exhibit very low magnetic hysteresis, therefore the domain wall motion takes place in low field and does not affect strongly the calculated anisotropy.

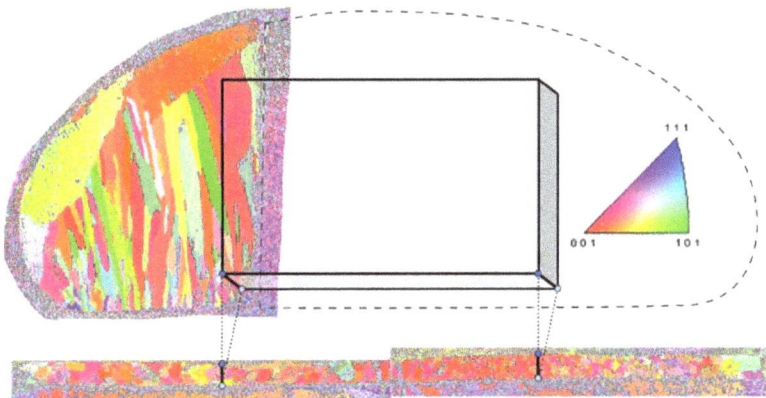

Figure 2. Columnar growth in arc-melted samples. The EBSD method demonstrates strong [001] texture. The sample for magnetic measurement was cut as shown. The bottom part shows predominant [001] orientation of the grains and its relation to the sample. The sample size is sketched and scaled proportionally to the EBSD images.

As the feasibility check we measured stoichiometric Ni$_2$MnGa. The magnetic energy determined from magnetization curve of polycrystalline material gives 2.8×10^5 J/m^3 comparable with the magnetocrystalline anisotropy of a single crystalline bulk 3.4–3.8×10^5 J/m^3 [9,30]. The measured value was smaller as expected from above discussion. The reasonable agreement, however, demonstrates that the method is applicable for the textured samples. In contrast, very low magnetocrystalline anisotropy of cubic austenite cannot be determined precisely and measured value is higher due to the effect of grains, imprecise correction for demagnetization field and other irregularities of polycrystals. Using our method, we determined the magnetic anisotropy of cubic stoichiometric Ni$_2$MnGa at room temperature to be about 20 kJ/m^3 which was an order of magnitude lower than for the martensite but somehow higher than value determined from single crystal [7,31].

Two groups of alloyed Ni-Mn-Ga were measured. To evaluate the effect of transitional metal alloying in a systematic way, we substituted 5% of each element in stoichiometric alloy with transitional metal Ni, Fe or Co [28]. Nine compositions were prepared in which one was stoichiometric. The anisotropy values for all compositions are summarized in Figure 3. The maximum anisotropy appeared for stoichiometric alloy and the anisotropy strongly depended not only on substituting element but also what element was substituted. The substitution of Ni by Fe resulted in cubic (austenite) structure down to at least 10 K. In this case, the determined magnetic energy was the order of magnitude smaller and comparable to stoichiometric Ni$_2$MnGa. In martensites, the differences in the anisotropy were significant, well above the expected error. It is apparent that adding Co to the alloy strongly decreases the magnetocrystalline anisotropy and more so if Ga is substituted. The decrease of anisotropy by Co alloying is in line with theoretical prediction [20] and experiment [30]. The decrease of anisotropy agreed with [25] which shows that anisotropy decreases more than twice for 5 at. % Gd or Ti substitution of Ga.

The other set is concerning with small amount of alloying by Fe and Cu to slightly off-stoichiometric alloys in attempt to further increase transformation temperatures [29]. For all these alloys, Table 1 summarizes the effective magnetocrystalline anisotropy determined from the magnetization loops. The anisotropy, as a function of the alloying Fe and Cu elements, is also drawn in Figure 4. Apparently, there is no correlation between amount of alloying and anisotropy. Although the small amount of Fe and Cu have a significant effect on martensite transformation temperature, it does not affect the magnetocrystalline anisotropy in any systematic way. From our data, it is difficult to judge the effect of small

Cu and Fe alloying. For Mn = 25 at. %, the anisotropy slightly increases with increasing Cu content; however, for 27.5 at% of Mn the tendency is just the opposite (Table 1).

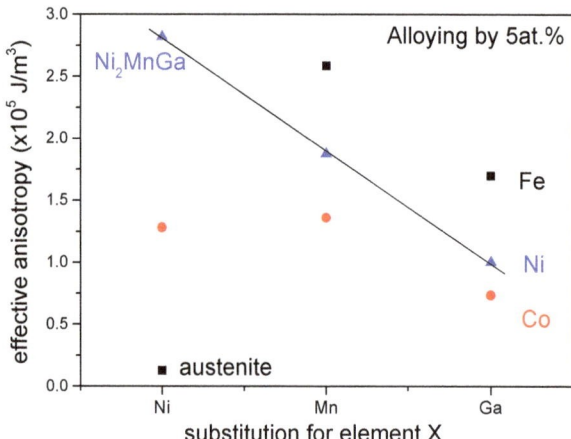

Figure 3. Effective magnetocrystalline anisotropy of Ni$_2$MnGa substituted by 5 at. % of Fe, Co and Ni on different position of the compound, i.e., for Ni, Mn, and Ga as marked. The alloy Ni$_{45}$Fe$_5$Mn$_{25}$Ga$_{25}$ is in austenite state and the anisotropy is the order of magnitude lower and comparable to stoichiometric Ni$_2$MnGa austenite (see text). The line is just a guide for eyes.

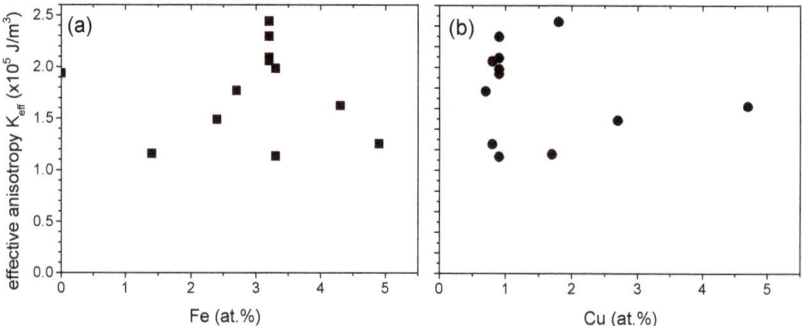

Figure 4. Effective magnetic anisotropy as a function of alloying elements (**a**) Fe (**b**) Cu in Ni-Mn-Ga.

In general, the alloying by Fe and Cu resulted in the anisotropy decrease in agreement with 5% set. As demonstrated in Figure 5, the anisotropy slightly decreased with increasing Mn content in agreement with Albertini [9] but there were analyzed only three compositions. Surprisingly, the slight decrease of Ni from stoichiometric 50% resulted mostly in the magnetic anisotropy increase. In agreement with the first set, the decrease of anisotropy occurred when Ga content decreased (Figure 5), i.e., when an alloying element substituted Ga.

These inconclusive results are difficult to interpret. The anisotropy fluctuation and missing expected trends may be ascribed to the imperfect ordering or disordering of the structure by alloyed elements. It is not known to which atomic positions the alloyed elements settle which can strongly affect the anisotropy values. In addition, possible different martensite structures appearing at 10 K can contribute to the scatter.

Despite unclear trends and tendencies, the findings have a practical impact for estimation of MIR effect in given alloys. The presented experimental results can also guide theoretical computation being able to predict new MSM materials, but it is also clear that further experimental investigation is needed.

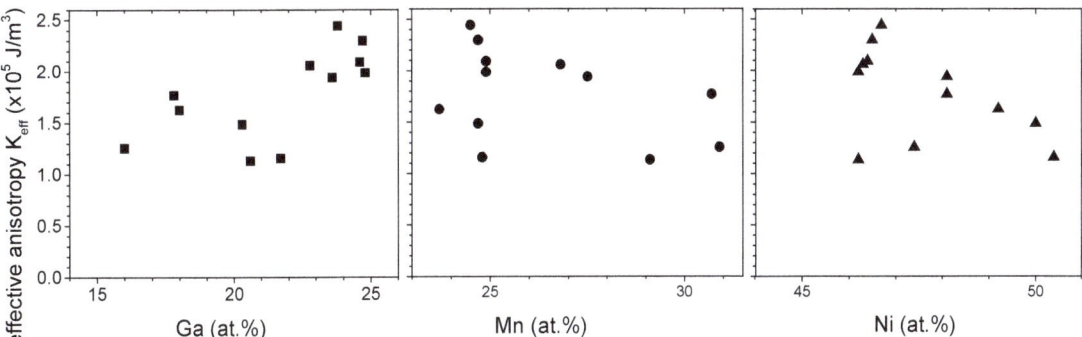

Figure 5. Effective magnetic anisotropy as a function of Ni, Mn and Ga content in Ni-Mn-Ga alloyed with Cu and Fe.

4. Conclusions

In Ni_2MnGa alloyed with 3d transition metals (Ni, Fe, Co) the magnetic anisotropy decreases. The maximum value of anisotropy is in stoichiometry. The strongest decrease is observed when alloyed element substitutes Ga.

Magnetocrystalline anisotropy in non-stoichiometric Ni-Mn-Ga decreases with several percent Fe and Cu alloying but no clear trend with increasing alloying is observed.

The decrease of anisotropy with alloying is weaker in contrast to calculations, which predicts even zero anisotropy for larger Cu and Co alloying.

Author Contributions: Conceptualization, O.H.; validation, M.R. and O.H.; formal analysis, M.R. and V.K.; investigation, M.R. and V.K.; writing—original draft preparation, M.R.; writing—review and editing, M.R., V.K. and O.H.; supervision, O.H.; funding acquisition, O.H. All authors have read and agreed to the published version of the manuscript.

Funding: This work was supported by Czech Science Foundation, grant No. 19-09882S. Magnetic measurements were performed in MGML (http://mgml.eu, accessed 12 December 2021), which is supported within the program of Czech Research Infrastructures (project no. LM2018096). V.K. would like to acknowledge financial support by the grant SGS19/190/OHK4/3T/14.

Institutional Review Board Statement: Not applicable.

Informed Consent Statement: Not applicable.

Data Availability Statement: Not applicable.

Acknowledgments: We thank Ladislav Klimša for providing EBSD analysis and Martin Dušák for alloys preparation.

Conflicts of Interest: The authors declare no conflict of interest. The funders had no role in the design of the study; in the collection, analyses, or interpretation of data; in the writing of the manuscript, or in the decision to publish the results.

References

1. Heczko, O. Magnetic shape memory effect and highly mobile twin boundaries. *Mater. Sci. Technol.* **2014**, *30*, 1559–1578. [CrossRef]
2. Ullakko, K.; Huang, J.K.; Kantner, C.; O'Handley, R.C.; Kokorin, V.V. Large magnetic-field-induced strains in Ni_2MnGa single crystals. *Appl. Phys. Lett.* **1996**, *69*, 1966–1968. [CrossRef]
3. Chikazumi, S. *Physics of Ferromagnetism*, 2nd ed.; Oxford University Press: Oxford, UK, 1997.
4. Ullakko, K.; Huang, J.K.; Kokorin, V.V.; O'Handley, R.C. Magnetically controlled shape memory effect in Ni_2MnGa intermetallics. *Scr. Mater.* **1997**, *36*, 1133–1138. [CrossRef]
5. Heczko, O.; Sozinov, A.; Ullakko, K. Giant field-induced reversible strain in magnetic shape memory NiMnGa alloy. *IEEE Trans. Magn.* **2000**, *36*, 3266–3268. [CrossRef]
6. Likhachev, A.A.; Ullakko, K. Quantitative model of large magnetostrain effect in ferromagnetic shapell memory alloys. *Eur. Phys. J. B* **2000**, *14*, 263–267. [CrossRef]

7. Tickle, R.; James, R.D. Magnetic and magnetomechanical properties of Ni$_2$MnGa. *J. Magn. Magn. Mater.* **1999**, *195*, 627–638. [CrossRef]
8. Straka, L.; Heczko, O.; Ullakko, K. Investigation of magnetic anisotropy of Ni-Mn-Ga seven-layered orthorhombic martensite. *J. Magn. Magn. Mater.* **2004**, *272*, 2049–2050. [CrossRef]
9. Albertini, F.; Pareti, L.; Paoluzi, A.; Morellon, L.; Algarabel, P.A.; Ibarra, M.R.; Righi, L. Composition and temperature dependence of the magnetocrystalline anisotropy in Ni$_{2+x}$Mn$_{1+y}$Ga$_{1+z}$ (x+y+z=0) Heusler alloys. *Appl. Phys. Lett.* **2002**, *81*, 4032–4034. [CrossRef]
10. Koho, K.; Söderberg, O.; Lanska, N.; Ge, Y.; Liu, X.; Straka, L.; Vimpari, J.; Heczko, O.; Lindroos, V.K. Effect of the chemical composition to martensitic transformation in Ni-Mn-Ga-Fe alloys. *Mater. Sci. Engin. A* **2004**, *378*, 384–388. [CrossRef]
11. Karaman, I.; Karaca, H.E.; Basaran, B.; Lagoudas, D.C.; Chumlyakov, Y.I.; Maier, H.J. Stress-assisted reversible magnetic field-induced phase transformation in Ni$_2$MnGa magnetic shape memory alloys. *Scr. Mater.* **2006**, *55*, 403–406. [CrossRef]
12. Gaitzsch, U.; Potschke, M.; Roth, S.; Rellinghaus, B.; Schultz, L. A 1% magnetostrain in polycrystalline 5M Ni-Mn-Ga. *Acta Mater.* **2009**, *57*, 365–370. [CrossRef]
13. Soto-Parra, D.; Moya, X.; Mañosa, L.; Planes, A.; Flores-Zúñiga, H.; Alvarado-Hernández, F.; Ochoa-Gamboa, R.; Matutes-Aquino, J.; Ríos-Jara, D. Fe and Co selective substitution in Ni$_2$MnGa: Effect of magnetism on relative phase stability. *Philos. Mag.* **2010**, *90*, 2771–2792. [CrossRef]
14. Kanomata, T.; Kitsunai, Y.; Sano, K.; Furutani, Y.; Nishihara, H.; Umetsu, R.Y.; Kainuma, R.; Miura, Y.; Shirai, M. Magnetic properties of quaternary Heusler alloys Ni$_{2-x}$Co$_x$MnGa. *Phys. Rev. B* **2009**, *80*, 214402. [CrossRef]
15. Sakon, T.; Fujimoto, N.; Kanomata, T.; Adachi, Y. Magnetostriction of Ni$_2$Mn$_{1-x}$Cr$_x$Ga Heusler Alloys. *Metals* **2017**, *7*, 410. [CrossRef]
16. Gomes, A.M.; Khan, M.; Stadler, S.; Ali, N.; Dubenko, I.; Takeuchi, A.Y.; Guimarães, A.P. Magnetocaloric properties of Ni$_2$Mn$_{1-x}$(Cu,Co)$_x$Ga Heusler alloys. *J. Appl. Phys.* **2006**, *99*, 08Q106. [CrossRef]
17. Adachi, Y.; Kouta, R.; Fujio, M.; Kanomata, T.; Umetsu, R.Y.; Xu, X.; Kainuma, R. Magnetic Phase Diagram of Heusler alloy system Ni$_2$Mn$_{1-x}$Cr$_x$Ga. *Phys. Proc.* **2015**, *75*, 1187–1191. [CrossRef]
18. Bozorth, R.M. *Ferromagnetism*; D. van Nostrand Company: New York, NY, USA, 1951.
19. Heczko, O.; Straka, L.; Novak, V.; Fähler, S. Magnetic anisotropy of nonmodulated Ni-Mn-Ga martensite revisited. *J. Appl. Phys.* **2010**, *107*, 09A914. [CrossRef]
20. Zelený, M.; Straka, L.; Rameš, M.; Sozinov, A.; Heczko, O. Origin of magnetocrystalline anisotropy in Ni-Mn-Ga-Co-Cu tetragonal martensite. *J. Magn. Magn. Mater.* **2020**, *503*, 166522. [CrossRef]
21. Albertini, F.; Solzi, M.; Paoluzi, A.; Righi, L. Magnetocaloric Properties and Magnetic Anisotropy by Tailoring Phase Transitions in NiMnGa Alloys. *Mater. Sci. Forum* **2008**, *583*, 169–196. [CrossRef]
22. Gavriljuk, V.G.; Dobrinsky, A.; Shanina, B.D.; Kolesnik, S.P. A study of the magnetic resonance in a single-crystal Ni$_{50.47}$Mn$_{28.17}$Ga$_{21.36}$ alloy. *J. Phys.–Condes. Matter* **2006**, *18*, 7613–7627. [CrossRef]
23. Kraus, L.; Heczko, O. Magnetic order in Mn excess Ni-Mn-Ga Heusler alloy single crystal probed by ferromagnetic resonance. *J. Magn. Magn. Mater.* **2021**, *532*, 167983. [CrossRef]
24. Chernenko, V.A.; Lvov, V.A.; Golub, V.; Aseguinolaza, I.R.; Barandiaran, J.M. Magnetic anisotropy of mesoscale-twinned Ni-Mn-Ga thin films. *Phys. Rev. B* **2011**, *84*, 054450. [CrossRef]
25. Łaszcz, A.; Hasiak, M.; Kaleta, J. Temperature Dependence of Anisotropy in Ti and Gd Doped NiMnGa-Based Multifunctional Ferromagnetic Shape Memory Alloys. *Materials* **2020**, *13*, 2906. [CrossRef] [PubMed]
26. Cejpek, P.; Straka, L.; Veis, M.; Colman, R.; Dopita, M.; Holý, V.; Heczko, O. Rapid floating zone growth of Ni$_2$MnGa single crystals exhibiting magnetic shape memory functionality. *J. Alloy. Compd.* **2019**, *775*, 533–541. [CrossRef]
27. Enkovaara, J.; Ayuela, A.; Nordström, L.; Nieminen, R.M. Magnetic anisotropy in Ni$_2$MnGa. *Phys. Rev. B* **2002**, *65*, 134422. [CrossRef]
28. Kopecký, V.; Rameš, M.; Veřtát, P.; Colman, R.H.; Heczko, O. Full Variation of Site Substitution in Ni-Mn-Ga by Ferromagnetic Transition Metals. *Metals* **2021**, *11*, 850. [CrossRef]
29. Armstrong, A.; Nilsén, F.; Rameš, M.; Colman, R.H.; Veřtát, P.; Kmječ, T.; Straka, L.; Müllner, P.; Heczko, O. Systematic Trends of Transformation Temperatures and Crystal Structure of Ni-Mn-Ga-Fe-Cu Alloys. *Shap. Mem. Superelast.* **2020**, *6*, 97–106. [CrossRef]
30. Rameš, M.; Heczko, O.; Sozinov, A.; Ullakko, K.; Straka, L. Magnetic properties of Ni-Mn-Ga-Co-Cu tetragonal martensites exhibiting magnetic shape memory effect. *Scr. Mater.* **2018**, *142*, 61–65. [CrossRef]
31. Heczko, O.; Kopeček, J.; Majtás, D.; Landa, M. Magnetic and magnetoelastic properties of Ni-Mn-Ga—Do they need a revision? *J. Phys. Conf. Ser.* **2011**, *303*, 012081. [CrossRef]

Article

Neutron Diffraction Study of the Martensitic Transformation of $Ni_{2.07}Mn_{0.93}Ga$ Heusler Alloy

Lara Righi [1,2]

[1] Department of Chemistry, Life Sciences and Environmental Sustainability, University of Parma, Parco Area delle Scienze 17/A, 43124 Parma, Italy; lara.righi@unipr.it
[2] IMEM-CNR, Parco Area delle Scienze 37/A, 43124 Parma, Italy

Citation: Righi, L. Neutron Diffraction Study of the Martensitic Transformation of $Ni_{2.07}Mn_{0.93}Ga$ Heusler Alloy. *Metals* **2021**, *11*, 1749. https://doi.org/10.3390/met11111749

Academic Editor: Enrique Jimenez-Melero

Received: 29 September 2021
Accepted: 27 October 2021
Published: 31 October 2021

Publisher's Note: MDPI stays neutral with regard to jurisdictional claims in published maps and institutional affiliations.

Copyright: © 2021 by the author. Licensee MDPI, Basel, Switzerland. This article is an open access article distributed under the terms and conditions of the Creative Commons Attribution (CC BY) license (https://creativecommons.org/licenses/by/4.0/).

Abstract: The martensitic transition featuring the ternary Heusler alloy $Ni2.09Mn0.91Ga$ was investigated by neutron diffraction. Differential scanning calorimetry indicated that structural transition starts at 230 K on cooling with a significant increase in the martensitic transformation onset compared to the classical Ni_2MnGa. The low-temperature martensite presents the 5M type of modulated structure, and the structural analysis was performed by the application of the superspace approach. As already observed in Mn-rich modulated martensites, the periodical distortion corresponds to an incommensurate wave-like shift of the atomic layers. The symmetry of the modulated martensite at 220 K is orthorhombic with unit cell constants a = 4.2172(5) Å, b = 5.5482(2) Å, and c = 4.1899(2) Å; space group $Immm(00\gamma)s00$; and modulation vector $q = \gamma c^* = 0.4226(5)c^*$. Considering the different neutron scattering lengths of the elements involved in this alloy, it was possible to ascertain that the chemical composition was $Ni_{2.07}Mn_{0.93}Ga$, close to the nominal formula. In order to characterize the martensitic transformation upon increasing the temperature, a series of neutron diffraction patterns was collected at different temperatures. The structural analysis indicated that the progressive change of the martensitic lattice is characterized by the exponential change of the c/a parameter approaching the limit value c/a = 1 of the cubic austenite.

Keywords: Heusler alloys; martensitic transformation; neutron diffraction

1. Introduction

Ni-Mn-X with X = Ga, In, Sn, Sb Heusler alloys are considered the prototypical multifunctional material encompassing a huge magnetocaloric effect (MCE) [1], magnetic field giant induced strains (MFIS), magnetic shape memory effect, and interesting exchange bias phenomena [2–4]. Most of the stunning physical properties of this class of intermetallics are based on the martensitic phase transformation (MT) taking place by temperature change [5,6], application of mechanical loads [7,8], or induced by the application of an external magnetic field [3,4]. In Heusler alloys, the parent phase exhibits the $L2_1$ superstructure, and the martensitic transition is accompanied by a lattice distortion yielding tetragonal, orthorhombic, or monoclinic structures (see Figure 1) depending on the chemical composition [6,7]. In the case of Ni-Mn based alloys, it is frequently observed that the martensitic phases are featured by modulated structural distortions generally associated to five-layered or seven-layered supercells defined 5M or 7M, respectively [7,9]. In Ni_2MnGa, the martensitic phase stable below 200 K is characterized by an incommensurate modulated 5M-type crystal structure that was successfully solved within the crystallographic approach called superspace [9,10]. Indeed, the MCE is related to the drastic entropy change that can be induced by a magnetic field change during the martensitic transformation or eventually at the corresponding Curie temperature [11].

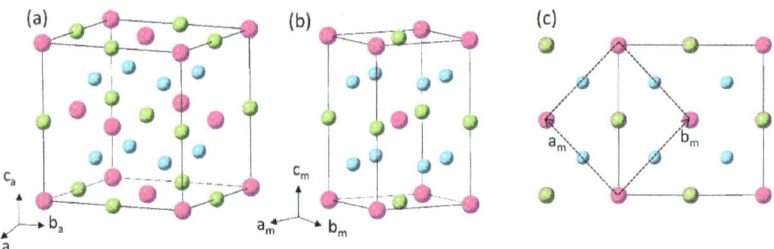

Figure 1. Fundamental crystal structures of austenitic (**a**) and martensitic (**b**) phases wherein the colored spheres define different crystallographic sites (light green = Ga, cyan = Ni, pink = Mn). (**c**) Reciprocal relation between the two lattices. The a and m subscripts of the crystallographic axes stand for austenite and martensite respectively.

In this context, the compositional combination $Ni_{2+x}Mn_{1-x}Ga$ has received increasing attention [12]. Large magnetocaloric effects in single crystals of $Ni_{2+x}Mn_{1-x}Ga$ [13] alloy are observed upon application of magnetic fields suitable for diffuse technological applications. A recently published study is focused on the intermartensitic transformation occurring for the variant $Ni_{2.15}Mn_{0.85}Ga$ [14]. The authors described a structural transition occurring at 215 K from the low-temperature tetragonal $L1_0$ phase to monoclinic modulated 7M-type martensite. Furthermore, the martensitic 7M phase is stable at room temperature, and the transition from martensite to austenite is reported to take place beyond 325 K. A further increment of Ni to reach $Ni_{2.18}Mn_{0.82}Ga$ composition stabilizes the tetragonal $L1_0$ martensite at room temperature.

Moreover, in such a Ni-rich alloy both the MT and Curie temperature merge in a magnetostructural first-order transition [15,16] with a beneficial effect on the MCE outcome. Besides compositions showing $0.15 \leq x \leq 0.18$, the compositional regimes based on $x < 0.1$ include valuable magnetic and MCE properties. Single crystals with $Ni_{2.04}Mn_{0.96}Ga$ composition were investigated by T. Liang [17] for the giant strains induced by external magnetic field. Strain-temperature curves evidenced that the martensitic phase undergoes below 240 K on cooling. Moreover, it was demonstrated that off-stoichiometric $Ni_{2.08}Mn_{1.04}Ga_{0.88}$ alloy shows interesting magnetocaloric properties in polycrystalline ribbons [14]. Notably, in a $Ni_{2+x}Mn_{1-x}Ga$ system, the compositions showing a moderate excess of Ni are still not completely explored, and structural studies of the martensitic transitions have not yet been reported. In this work, the structural investigation of the nominal $Ni_{2.09}Mn_{0.98}Ga$ Heusler alloy was undertaken by neutron diffraction experiments. It was found that, below 230 K, the martensitic phase displays an incommensurate 5M modulated crystal structure. The structural refinement based on the neutron diffraction data was achieved by employing the model obtained for the Ni_2MnGa counterpart. Furthermore, the study of the structural transformation versus temperature points out that the progressive change of the martensitic lattice is closely related to geometrical constraints in agreement with the invariant plane strain requirements.

2. Materials and Methods

The sample with nominal composition $Ni_{2.09}Mn_{0.91}Ga$ was obtained from high purity elements by arc melting and subsequent annealing in a quartz ampoule under Ar atmosphere and quenched in cold water. The annealing treatment was performed at 983 K for 72 h. Afterwards, the small ingots were ground in a mortar and successively annealed at 1073 K for 2 h in order to remove the defects induced by the grinding process. Powder X-ray diffraction measurement, performed with an X'TRA Thermo diffractometer (Thermo Fisher Scientific, Losanna, Switzerland) equipped with a CuKα source and a solid-state detector, indicated the presence of single-phase austenite (see Figure S1 in the Supplementary Information).

Differential scanning calorimetry (DSC) was carried out under nitrogen in the temperature range from room temperature to 200 K and then upon heating from 200 K to 350 K. The cooling and heating scans were performed with a rate of 10 K/min

Neutron diffraction experiments were performed on the D2B beamline (ILL, Grenoble, France) at ILL (Institute Laue-Longevin, Grenoble, France) with a wavelength of 1.594 Å in the temperature range from 220 K to 350 K, and diffraction patterns were collected at selected temperatures. The low-temperature diffraction pattern (220 K) used to determine the crystal structure of the martensitic phase was recorded at a low scanning rate in order to assure the required statistics and a good peak/background ratio. The structural refinements were based on the Rietveld method [18] with the JANA2006 suite (Institute of Physics Department of Structure Analysis, Praha, Czech Republic).

3. Results and Discussion

The critical temperature of the martensitic transition for the prepared Ni-rich composition was determined by thermal analysis. The DSC measurement, illustrated in Figure 2a, manifested a pronounced transformation at 231 K on cooling, and the reverse martensitic transformation is observed at 244 K. The minute substitution of Mn with Ni in the stoichiometric Ni_2MnGa alloy induces an increase of 35 K at the MT onset. Indeed, the critical temperatures recorded by DSC scans are in agreement with those reported for a similar Ni-rich composition [17].

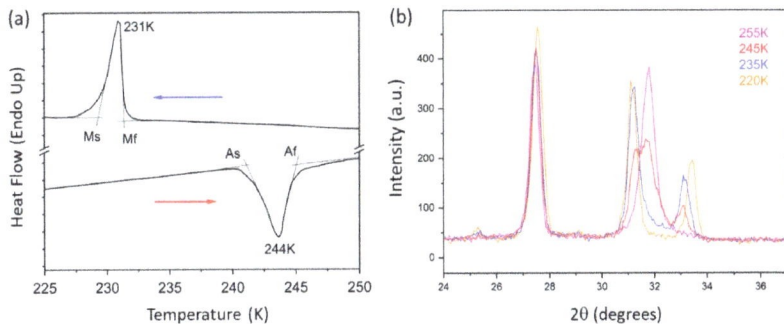

Figure 2. (a) DSC scan of the $Ni_{2.07}Mn_{0.93}Ga$ sample (Ms = start of martensite, Mf = finish of martensite, As = start of austenite, Af = finish of austenite). (b) Neutron diffraction patterns in the 24–37° 2θ range collected at different temperatures.

A neutron diffraction experiment on the polycrystalline alloy with the nominal formula $Ni_{2.09}Mn_{0.91}Ga$ was undertaken to establish the martensitic structure and the structural evolution of the martensitic transformation. In Figure 2b, a limited region of the neutron diffraction patterns collected at different increasing temperatures is shown. In accordance with the indication provided by DSC, the martensitic transition takes place beyond 240 K.

The phase diagram accounting for the T_c and MT in the $Ni_{2+x}Mn_{1-x}Ga$ system and reported by V.V. Kovaylo et al. [12] indicates that for a moderate excess of Ni, the paramagnetic state occurs above 350 K. Therefore, a neutron diffraction collection was performed at 350 K above T_c in order to suppress the magnetic contribution to the scattering and determine the chemical composition of the austenitic phase. Neutron diffraction data are particularly suited for the study of the chemical order in the Heusler lattice. The neutron scattering lengths are not Z-dependent as in the case of X-ray diffraction [19], and this allows the faithful determination of occupancy factors for each crystallographic site from powder diffraction experiments. Neutron diffraction studies [20] dealing with the characterization of Heusler alloys have demonstrated that a Rietveld refinement is a suitable tool for the reliable determination of the stoichiometric composition of the prepared alloys. In the case of this Ni-rich alloy, the structural refinement was based on the cubic structure showing the Fm-3m space group typical for the L2$_1$ superstructure. The crystal data obtained by the structural refinement are listed in the Tables S1 and S2 reported in SI. The structural model was conceived by considering an excess of Ni in the site of Mn; this assumption is in agreement with previous studies [15–17] that accounted for the predominant tendency of the Ni excess to occupy the Mn position. The chemical ratio of Ni and Mn located in the same site was constrained to maintaining the global occupancy of the site equivalent to 1. The best fitting (see Figure S2) was obtained with a cubic structure with the unit cell constant a = 5.83194(4) Å and occupancy factors of 0.93(5) and 0.07(5) for Mn and Ni, respectively, in the 2a site, yielding a chemical composition of the austenitic phase corresponding to Ni$_{2.07}$Mn$_{0.93}$Ga showing a slight deviation from the nominal formula. Since the martensitic transition is a diffusion-less process, the actual atomic distribution in the ordered lattice is preserved in the martensitic phase as well. The neutron diffraction pattern collected at 220 K shows a close analogy with that reported for the stoichiometric Ni$_2$MnGa compound [21]. As remarked before, below 200 K, the martensitic phase of Ni$_2$MnGa possesses an incommensurate modulated structure that was successfully solved with the application of the superspace approach [22]. The superspace approach has been recurrently employed to analyze the 5M and 7M modulated structures of martensites in Ni-Mn-based Heusler alloys [9,10,23–25]. In agreement with the crystallographic representation of these complex structures [21], the diffraction peaks are divided into two groups: the main reflections are related to the basic body-centered martensitic lattice depicted in Figure 1b, and the weak peaks called satellites are associated with the modulation vector $q = \alpha a^* + \beta b^* + \gamma c^*$ connected to the incommensurate distortion of the fundamental crystal structure. To perform the Rietveld refinement of the neutron diffraction pattern collected at 220 K, the orthorhombic model derived for 5M modulated martensite in Ni$_2$MnGa Heusler compound [9,10] was adopted. The basic martensitic structure (see Figure 1) originated by the Bain distortion of the austenitic L2$_1$ lattice is characterized by the body-centered Immm symmetry.

The (3+1)-dimensional structure belongs to the Immm(00γ)s00 superspace group and the modulated atomic positions of Ni Mn and Ga are defined by a wave-like shape expressed by a sinusoidal function:

$$u^i(x_4) = A_1^i \sin(2\pi n x_4) \qquad (1)$$

that is based on the x_4 superspace coordinate for j-th atomic site and the n index representing the order of the Fourier series [21]. Further details of the superspace approach are illustrated in the Supplementary Information. Therefore, the refinement of the amplitudes A_1^i allows to find the incommensurate modulated displacing of the atomic positions. Main reflections and satellites are used in the Rietveld refinement and convergence is achieved for the values summarized in Tables S3 and S4. The refined modulation vector $q = \gamma c^*$ (α and β variables assume zero value for symmetry) exhibits a deviation from the commensurate value of 0.4 c^* = 2/5 c^*; indeed, in analogy with what was already found for the stoichiometric Ni$_2$MnGa [9], this martensite can be regarded as incommensurate modulated 5M having q = 0.4226 c^* = 2/5(1 + δ)c^* where δ = 0.023 defines the aperiodicity of the system. Figure 3a represents the wave-like shift of the atomic layers stacked along the c axis where the atomic coordinates are extrapolated by assuming an ideal q = 2/5c^* commensurate condition. In Figure 3b, the incommensurate deviation from the fundamental positions in the basic lattice is depicted. For all three atomic sites, the amplitudes A1x assume rather similar values, indicating that the systematic shift is damped to the 00l planes. Figure 3c reports the Rietveld plot of the structural refinement with agreement factors R_p = 6.45% and R_{wp} = 9.23%.

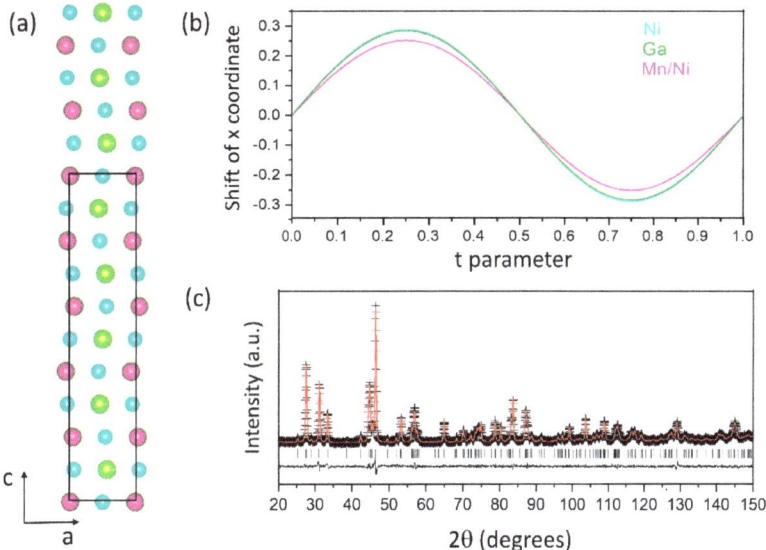

Figure 3. (a) View of the incommensurate modulated structure of the $Ni_{2.07}Mn_{0.93}Ga$ martensitic phase. (b) Sinusoidal displacement of the atomic positions obtained by the Rietveld refinement of the (3+1)-dimensional model. The t parameter is related to the x_4 coordinate. (c) Rietveld plot of the structural refinement for the modulated martensite at 220 K.

The magnetic characterizations involving $Ni_{2+x}Mn_{1-x}Ga$ alloys with a slight excess of Ni revealed ferromagnetism in the low temperature martensitic phase [12]. Therefore, the structural refinement was carried out considering a ferromagnetic alignment of the spins along the b axis of the orthorhombic martensitic lattice as illustrated in Figure S3. The structural refinement returns a magnetic moment located on the Mn site of 2.86(6) μ_B for the modulated martensite at 220 K. The decrement of the magnetic moment if compared to that reported for Mn in the classical Ni_2MnGa [26] is in agreement with magnetic measurements described in previous publications [12,27].

This effect regards the dilution of the Mn atoms distributed in the Heusler sublattice with Ni atoms that bring a small, localized moment (0.33 μ_B approximately [26]). The structural evolution of the martensitic phase across the structural transformation is monitored by collecting neutron diffraction patterns at selected temperatures from 220 K to 250 K. The values associated with each temperature were determined by changing the labelling of the crystallographic axes from abc to acb (with reference to the crystal data reported in Table S3; in this new set the martensitic lattice has the dimensions a_m = 4.2167 Å, b_m = 4.1870 Å, c_m = 5.5484 Å, and $\alpha = \beta = \gamma = 90°$) in order to assign the shorter unit cell constant along the c axis. From 220 K to 240 K, the modulated martensite manifests a slight change of the unit cell parameters as indicated in Figure 4a. The growth of the austenite takes place at around 240 K, and as illustrated in the inset of Figure 4a, this phase linearly increments its relative amount as the temperature rises. Beyond 240 K, the crystal lattice of the martensite displays a progressive change with a marked elongation of the shorter axis. By increasing the temperature, the parameter of the modulation vector shows a decrement of its value toward the 0.4 condition (see Figure 4b) equivalent to the commensurate variant of the 5M martensite [9]. This trend was observed for Mn-rich Ni-Mn-Ga martensitic 5M modulated structures [9,23], and it is a clear indication that the wave-like modulation is dominated by the phonon softening of TA_2 modes as remarked by Zheludev et al. [28]. The evolution of the tetragonality parameters is depicted in Figure 4c against the temperature change. The tetragonality is calculated considering a pseudo-tetragonal lattice with $a = \sqrt{2}a_m$ and $b = \sqrt{2}b_m$. Indeed, upon heating, the martensite manifests a progressive rearrangement of the lattice dimensions to mitigate the Bain distortion. This behavior is observed below the MT temperature but is particularly evident with the beginning of austenite growth. This evidence confirms that, in agreement with the crystallographic theory of martensitic transformations [29], as long as the structural transition from martensite to austenite is ongoing, the two coexisting crystal structures are metrically correlated to comply with the invariant plane strain requirements. The structural constraint forcing the martensitic unit cell parameters to assume certain values can be also studied by the application of the co-factor approach between 235 and 255 K.

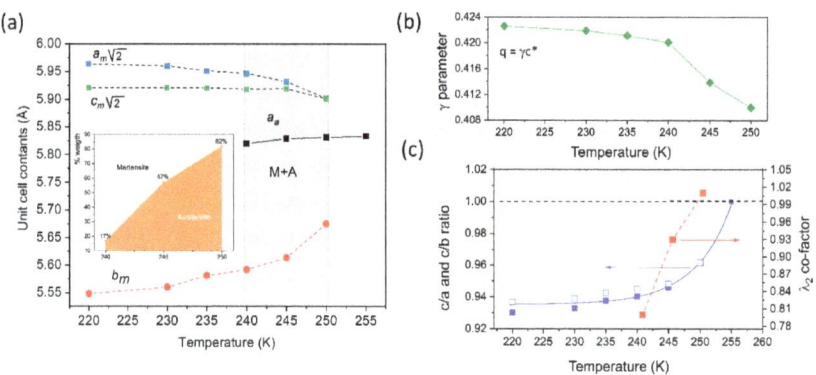

Figure 4. (**a**) Unit cell parameter change versus temperature. The a and b parameters are rescaled to the austenitic lattice. The grey region indicates the temperature range with coexistence of austenite and martensite. Inset: reciprocal amount of parent and product phase during the MT. (**b**) Temperature dependence of the γ variable of the modulation vector. (**c**) Tetragonality parameters c/a, c/b, and co-factor λ_2 against temperature.

The co-factor λ_2 was introduced by J.M. Ball and R.D. James [30,31] in order to evaluate the compatibility conditions of martensitic and austenitic lattice. A co-factor parameter λ_2 equal to 1 indicates zero elastic energy of the austenite/martensite interface, therefore implying a stress-free propagation of the phase boundaries during the MT process [30–34]. One of the implications of the condition $\lambda_2 = 1$ is the narrowing of the thermal hysteresis of the first-order martensitic transition. The temperature span of the direct and reverse martensitic transition is a detrimental aspect for the technological application of the MCE, and theoretical tools such as the co-factor method were adopted to predict a favorable compositional chemistry to control the thermal drift. Further, the co-factor was applied to the Ni-Mn-Ga Heusler alloy to evaluate the lattice compatibility between modulated 7M martensites and the corresponding parent phase [35]. In the case of $Ni_{2.07}Mn_{0.93}Ga$ the co-factor is calculated to provide additional information on the reciprocal correlation of the martensitic unit cell parameters ruled by the invariant plane strain conditions. The matrix selected for the co-factor extrapolation and the crystallographic assumptions used for this approach are summarized in the Supplementary Information. As indicated in Figure 4c, from 240 to 250 K, the λ_2 parameter moves from 0.8 to 1.01, indicating a linear dependence with the temperature. Interestingly, the co-factor approaches the ideal value when the austenite exceeds 80% in weight.

4. Conclusions

The actual structural study brings new insights into the crystal structure of the martensitic phase characterizing $Ni_{2+x}Mn_{1-x}Ga$ compositions with $x < 1$. The crystal structure of the martensitic phase $Ni_{2.07}Mn_{0.93}Ga$ was solved by using neutron diffraction data. At 220 K this phase displays an incommensurate modulated 5M structure with a modulation vector closely related to the Ni_2MnGa stoichiometric compound. The structural analysis of neutron diffraction patterns collected at different temperatures upon heating was used to study the progressive mitigation of the classical Bain distortion of the martensite. As expected, the parameters c/a and c/b tend to assume higher values as the temperature increases, indicating the progressive restoring of the cubic symmetry. The co-factor parameter, calculated in the temperature range of the hysteresis of the transformation, highlights that the maximal lattice compatibility between austenite and martensite is reached when the volume of the parent phase is predominant. Likewise, this approach suggests that the mechanism ruling the diffusion-less transformation from orthorhombic martensite to cubic austenite is apparently not influenced by the structural modulation.

Supplementary Materials: The following are available online at https://www.mdpi.com/article/10.3390/met11111749/s1, Figure S1: X-ray diffraction pattern of the polycrystalline $Ni_{2.07}Mn_{0.93}Ga$ alloy, Figure S2: Rietveld plot based on neutron diffraction data for the austenitic phase at 350 K. Figure S3: Ferromagnetic orientation of the spins located on the Mn/Ni site of the orthorhombic body centered lattice. Table S1: Crystal data of austenitic phase at 350 K, Table S2: Atomic coordinates, occupancy factors and thermal parameters for the cubic austenite at 350 K. Table S3: Crystal data of the modulated martensitic phase based on the structural refinement of the neutron diffraction data collected at 220K. Table S4: Atomic coordinate, occupancy factors, thermal parameters and modulation parameters A1x for the martensite at 220 K.

Funding: This research received no external funding.

Data Availability Statement: Data presented in this article are available at request from the corresponding author.

Acknowledgments: The author is grateful to R. Magnani for technical support during thermal measurements. The experimental and scientific support provided by ILL (Institute Laue-Langevin, France) during the neutron diffraction experiments performed at D2B beamline It is acknowledged.

Conflicts of Interest: The author declares no conflict of interest.

References

1. Franco, V.; Blázquez, J.S.; Ipus, J.J.; Law, J.Y.; Moreno-Ramírez, L.M.; Conde, A. Magnetocaloric ef-fect: From materials research to refrigeration devices. *Prog. Mater. Sci.* **2018**, *93*, 112–232. [CrossRef]
2. Khan, R.A.A.; Ghomashchi, R.; Xie, Z.; Chen, L. Ferromagnetic Shape Memory Heusler Materials: Synthesis, Microstructure Characterization and Magnetostructural Properties. *Materials* **2018**, *11*, 988. [CrossRef] [PubMed]
3. Acet, M.; Mañosa, L.; Planes, A. Handbook of Magnetic. *Materials* **2011**, *19*, 231–289.
4. Kainuma, R.; Imano, Y.; Ito, W. Magnetic-field-induced shape recovery by reverse phase transformation. *Nature* **2006**, *439*, 957–960. [CrossRef] [PubMed]
5. Pons, J.; Chernenko, V.A.; Santamarta, R.; Cesari, E. Crystal structure of martensitic phases in Ni–Mn–Ga shape memory alloys. *Acta Mater.* **2000**, *48*, 3027–3038. [CrossRef]
6. Chernenko, V.A.; Cesari, E.; Pons, J.; Segui, C. Phase Transformations in Rapidly Quenched Ni–Mn–Ga Alloys. *J. Mater. Res.* **2000**, *15*, 1496–1504. [CrossRef]
7. Martynov, V.V. X-ray diffraction study of thermally and stress-induced phase transformations in single crystalline Ni-Mn-Ga alloys. *J. Phys. IV* **1995**, *C8*, 91–99. [CrossRef]
8. Salazar, M.C.; Küchler, R.; Nayak, A.K.; Felser, C.; Nicklas, M. Uniaxial-stress tuned large mag-netic-shape-memory effect in Ni-Co-Mn-Sb Heusler alloys. *Appl. Phys. Lett.* **2017**, *110*, 071901. [CrossRef]
9. Righi, L.; Albertini, F.; Fabbrici, S.; Paoluzi, A. Crystal Structures of Modulated Martensitic Phases of FSM Heusler Alloys. *Mater. Sci. Forum* **2011**, *684*, 105–116. [CrossRef]
10. Righi, L.; Albertini, F.; Calestani, G.; Pareti, L.; Paoluzi, A.; Ritter, C.; Algarabel, P.A.; Morellon, L.; Ibarra, M.R. Incommensurate modulated structure of the ferromagnetic shape-memory Ni2MnGa martensite. *J. Solid State Chem.* **2006**, *179*, 3525–3533. [CrossRef]
11. Singh, S.; Caron, L.; D'Souza, S.W.; Fichtner, T.; Porcari, G.; Fabbrici, S.; Shekhar, C.; Chadov, S.; Solzi, M.; Felser, C. Large Magnetization and Reversible Magnetocaloric Effect at the Second-Order Magnetic Transition in Heusler. *Mater. Adv. Mater.* **2016**, *28*, 3321–3325. [CrossRef]
12. Khovaylo, V.V.; Buchelnikov, V.D.; Kainuma, R.; Koledov, V.V.; Ohtsuka, M.; Shavrov, V.G.; Takagi, T.; Taskaev, S.V.; Vasiliev, A.N. Phase transitions in Ni2+xMn1-xGa with a high Ni excess. *Phys. Rev. B* **2005**, *72*, 224408. [CrossRef]
13. Cherechukin, A.; Takagi, T.; Matsumoto, M.; Buchel'nikovc, V.D. Magnetocaloric effect in Ni2+xMn1−xGa Heusler alloys. *Phys Lett. A* **2004**, *326*, 146–151. [CrossRef]
14. Li, Z.; Zhang, Y.; Sánchez-Valdés, C.F.; Sánchez Llamazares, J.L.; Esling, C.; Zhao, X.; Zuo, L. Giant magnetocaloric effect in melt-spun Ni-Mn-Ga ribbons with magneto-multistructural transformation. *Appl. Phys. Lett.* **2014**, *104*, 044101. [CrossRef]

15. Khovaylo, V.V.; Skokov, K.P.; Taskaev, S.V.; Karpenkov, D.U.; Dilmieva, E.T.; Koledov, V.V.; Koshkid'ko, Y.S.; Shavrov, V.G.; Buchelnikov, V.D. Magnetocaloric properties of Ni2+xMn1-xGa with coupled magnetostructural phase transition. *J. Appl. Phys.* **2020**, *127*, 173903. [CrossRef]
16. Kamantsev, A.P.; Koledov, V.; Mashirov, A.V.; Dilmieva, E.T.; Shavrov, V.G.; Cwik, J.; Los, A.S.; Nizhankovskii, V.I.; Rogacki, K.; Tereshina, I.S.; et al. Magnetocaloric and thermomagnetic properties of Ni2.18Mn0.82Ga Heusler alloy in high magnetic fields up to 140 kOe. *J. Appl. Phys.* **2015**, *117*, 163903. [CrossRef]
17. Liang, T.; Chengbao, J.; Huibin, X. Temperature dependence of transformation strain and magnet-ic field-induced strain in Ni51Mn24Ga25 single crystal. *Mater. Sci. Eng. A* **2005**, *402*, 5–8. [CrossRef]
18. Young, R.A. *The Rietveld Method*; Oxford University Press: Oxford, UK, 1995.
19. Li, L.Y.; Mattei, G.S.; Li, Z.; Zheng, J.; Zhao, W.; Omenya, F.; Fang, C.; Li, W.; Jianyu Li, J.; Xie, Q.; et al. Extending the limits of powder diffraction analysis: Diffraction parameter space, occupancy defects, and atomic form factors. *Rev. Sci. Instrum.* **2018**, *89*, 093002.
20. Barandiaran, J.M.; Gutiérrez, J.; Lázpita, P.; Feuchtwanger, J. Neutron Diffraction Studies of Magnetic Shape Memory Alloys. *Mater. Sci. Forum* **2011**, *684*, 73–84. [CrossRef]
21. Kushida, H.; Fukuda, K.; Terai, T.; Fukuda, T.; Kakeshita, T.; Ohba, T.; Osakabe, T.; Kakurai, K.; Kato, K. Crystal structure of martensite and intermediate phases in Ni2MnGa studied by neutron diffraction. *Eur. Phys. J.* **2008**, *158*, 87–92. [CrossRef]
22. Van Smaalen, S. Incommensurate crystal structures. *Crystallogr. Rev.* **1995**, *4*, 79–202. [CrossRef]
23. Pramanick, A.; Wang, X.P.; An, K.; Stoica, A.D.; Yi, J.; Gai, Z.; Hoffmann, C.; Wang, X.L. Structural modulations and magnetic properties of off-stoichiometric Ni-Mn-Ga magnetic shape memory alloys. *Phys. Rev. B* **2012**, *85*, 144412. [CrossRef]
24. Devi, P.; Singh, S.; Dutta, B.; Manna, K.; D'Souza, S.W.; Ikeda, Y.; Suard, E.; Petricek, V.; Simon, P.; Werner, P.; et al. Adaptive modulation in the Ni2Mn1.4In0.6 magnetic shape-memory Heusler alloy. *Phys. Rev. B* **2018**, *97*, 224102. [CrossRef]
25. Singh, S.; Ziebeck, A.; Suard, E.; Rajput, P.; Bhardwaj, S.; Awasthi, A.M.; Barman, S.R. Modulated structure in the martensite phase of Ni1.8Pt0.2MnGa: A neutron diffraction study. *Appl. Phys. Lett.* **2012**, *101*, 171904. [CrossRef]
26. Lázpita, P.; Barandiarán, J.M.; Gutiérrez, J.; Feuchtwanger, J.; Chernenko, V.A.; Richard, M.L. Magnetic moment and chemical order in off-stoichiometric Ni–Mn–Ga ferromagnetic shape memory alloys. *New J. Phys.* **2011**, *13*, 033039. [CrossRef]
27. Ahuja, B.L.; Sharma, B.K.; Mathur, R.; Heda, N.L.; Itou, M.; Andrejczuk, M.; Sakurai, Y.; Chakrabarti, A.; Banik, S.; Awasthi, A.M.; et al. Magnetic Compton scattering study of Ni2+xMn1-xGa ferromagnetic shape-memory alloys. *Phys. Rev. B* **2007**, *75*, 134403. [CrossRef]
28. Zheludev, A.; Shapiro, S.M.; Wochner, P.; Tanner, L.E. Precursor effects and premartensitic transformation in Ni2MnGa. *Phys. Rev. B* **1996**, *54*, 15045. [CrossRef]
29. Airoldi, G. *Shape Memory Alloys: From Microstructure to Macroscopic Properties*; Bhattacharya, G., Airoldi, I., Müller, S., Miyazaki, S., Eds.; Trans Tech Publications: Bäch, Switzerland, 1997.
30. Ball, J.M.; Carstensen, C. Nonclassical austenite-martensite interfaces. *J. Phys. IV* **1997**, *7*, 35–40. [CrossRef]
31. Ball, J.M.; James, R.D. Fine phase mixtures as minimizers of energy. *Arch. Ration. Mech. Anal.* **1987**, *100*, 13–52. [CrossRef]
32. Della, F.P. On the cofactor conditions and further conditions of supercompatibility. *J. Mech. Phys. Solids* **2019**, *122*, 27–53. [CrossRef]
33. Chen, X.; Srivastava, V.; Dabade, V.; James, R.D. Study of the cofactor conditions: Conditions of supercompatibility between phases. *J. Mech. Phys. Solids* **2013**, *61*, 2566–2587. [CrossRef]
34. James, R.D.; Hane, K.F. Martensitic transformations and shape-memory materials. *Acta Mater.* **2000**, *48*, 197–222. [CrossRef]
35. Devi, P.; Salazar Mejia, C.; Ghorbani Zavareh, M.; Dubey, K.K.; Kushwaha, P.; Skourski, Y.; Felser., C.; Nicklas, M.; Singh, S. Improved magnetostructural and magnetocaloric reversibility in magnetic Ni-Mn-In shape memory Heusler alloy by optimizing the geometric compatibility condition. *Phys. Rev. Mater.* **2019**, *3*, 062401. [CrossRef]

Article

Actuating a Magnetic Shape Memory Element Locally with a Set of Coils

Andrew Armstrong and Peter Müllner *

Micron School of Materials Science and Engineering, Boise State University, 1910 University Ave, Boise, ID 83725, USA; andrewarmstrong858@u.boisestate.edu
* Correspondence: petermullner@boisestate.edu

Abstract: The local actuation of a magnetic shape memory (MSM) element as used in an MSM micropump is considered. This paper presents the difference between an electromagnetic driver and a driver that uses a rotating permanent magnet. For the magnetic field energy of the permanent magnetic drive, the element takes in a significant stray field. In a particular case, energy reduction was 12.7 mJ. For an electromagnetic drive with an identical size of the MSM element, the total magnetic field energy created by the system was 2.28 mJ. Attempts to experimentally nucleate twins in an MSM element by energizing an electromagnetic drive failed even though the local magnetic field exceeded the magnetic switching field. The energy variation is an order of magnitude smaller for the electromagnetic drive, and it does not generate the necessary driving force. It was assumed in previous work that the so-called magnetic switching field presents a sufficient requirement to nucleate a twin and, thus, to locally actuate an MSM element. Here, we show that the total magnetic field energy available to the MSM element presents another requirement.

Keywords: Ni-Mn-Ga; FSMA; electromagnetic coupling; smart materials; actuator

---

Citation: Armstrong, A.; Müllner, P. Actuating a Magnetic Shape Memory Element Locally with a Set of Coils. *Metals* **2021**, *11*, 536. https://doi.org/10.3390/met11040536

Academic Editor: Ryosuke Kainuma

Received: 23 February 2021
Accepted: 19 March 2021
Published: 25 March 2021

Publisher's Note: MDPI stays neutral with regard to jurisdictional claims in published maps and institutional affiliations.

Copyright: © 2021 by the authors. Licensee MDPI, Basel, Switzerland. This article is an open access article distributed under the terms and conditions of the Creative Commons Attribution (CC BY) license (https://creativecommons.org/licenses/by/4.0/).

1. Introduction

Progress in modern technology depends upon novel materials with specific physical and functional properties. Some intermetallic alloys exhibit the shape memory effect, where the material experiences a shape change from an external stimulus such as temperature, magnetic field, and mechanical stress (e.g., [1–3]). Magnetic shape memory (MSM) alloys exhibit shape change resulting from magnetic field. Single crystal MSM alloy exhibits a large stroke and short actuation time with the ability to change its size and shape several million times. Essentially, the material operates as a metallic muscle controlled by the variation of a magnetic field. Ni-Mn-Ga exhibits a martensite phase with highly mobile twin boundaries. Upon the application of mechanical stress or a magnetic field, the twin domains reorient and enable high magnetic-field-induced strain. MSM alloys are considered for applications in microactuators, strain sensors [4], energy harvesters [5,6], and micropumps [7–9].

In a uniform magnetic field, the MSM element deforms by extending and contracting uniformly in the bulk, as would a piezoelectric element [10]. In contrast, a localized magnetic field causes a localized shrinkage in the MSM element, which can be moved along with the localized magnetic field [7]. Rotation of the localized field moves a shrinkage through an MSM element [11]. The motion of the shrinkage along the element can be used to build a pump in a similar manner to the esophageal contractions that mammals use to swallow food [7].

In previous work, the rotating magnetic field has been provided by the rotation of a diametrically magnetized permanent magnet, where a micromotor spun the magnet. Micromotors are large in comparison to an MSM sample and have moving parts prone to failure. In lieu of actuating the shrinkage with a physically rotating magnet, this study considers the advancement of a localized magnetic field using a miniature electromagnet with a plurality of magnetic poles. Previously, to consider the actuation of an MSM sample by localized electromagnetic coils, Smith et al. physically repositioned the poles of an electromagnet [12]. The authors found that a locally twinned region formed at the magnetic pole tips. In [13], we created a motionless magnetic driver in a device with multiple magnetic poles arranged in a row. Individual poles were energized with a strong magnetic field by compressing a magnetic field through two coils in opposition adjacent to a soft ferromagnetic pole. Changing the polarity of the coils energized other poles and advanced a magnetic field. The shrinkage formed on the MSM element near the pole tips. However, the spacing between poles was too coarse to move the deformed region.

In this paper, we present an electromagnetic drive with two rows of magnetic poles, similar to linear motor yokes, which are staggered across the MSM element. Each pole can be energized individually and sequentially to approximate a moving vertical field along the MSM element. A device was built to test the effect of the pulsing electromagnetic drive upon an MSM element. Along with physical results, magnetic models using finite element method magnetics (FEMM) were created. The material and magnetic behavior is compared to that known when driven by a rotating rare earth Nd-Fe-B magnet.

2. Materials and Methods

2.1. Device Design

The magnetic yokes were machined as shown in Figure 1 out of a 3.0 mm thick plate of Fe-Co (Vacoflux 50). The material is magnetically soft and supports 2.3 T at saturation [14]. The yoke was composed of a top and bottom yoke that interfaced via a friction fit. The bottom yoke had three poles and the top yoke had four poles. The bottom yoke slid into the top yoke. The two yokes had poles juxtaposed to each other. In Figure 1a, the yokes are separated by an air gap. The poles (P1–P7) were magnetized by the bottom coils, B1–B4, and the top coils T1–T5. Figure 1a shows the flux pattern when P1 and P2 are actuated with magnetic flux flowing upward.

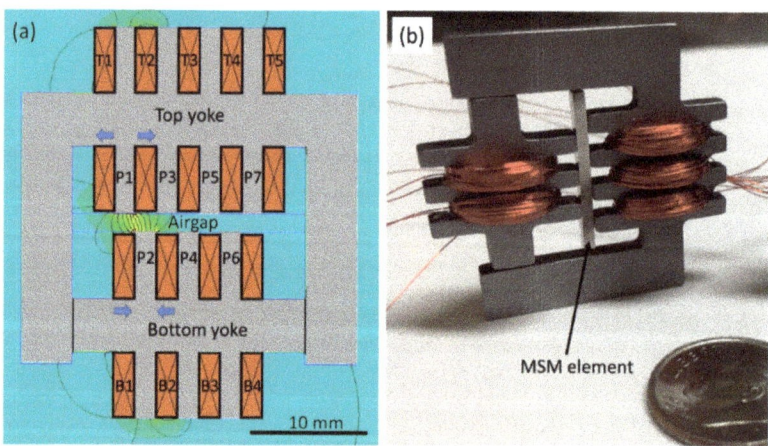

Figure 1. Design of the electromagnetic drive: (**a**) finite element method magnetics (FEMM) simulation including the Fe-Co yoke geometry, pole geometry, and coil geometry. The bottom yoke contains the bottom poles P2, P4, and P6 and slides between the tines of the top yoke. The bottom and top yokes had poles that were staggered relative to each other. (**b**) shows the machined device before winding the outermost coils.

Each side of the yoke had an inner row of poles and an outer row of poles. The outer poles were simply for coil containment when winding. Upon actuation, the magnetic circuit guided the vast majority of magnetic flux through the inner poles. The inner width of the yoke was 20 mm, which was designed to accommodate an MSM element. The edges were deburred by sandblasting such that coils could be wound directly onto the yokes. The coil wire had a polyamide coated and a diameter of 0.13 mm (36 AWG). Figure 1b shows the yokes with only five coils, while for the experiments, four coils were added outside of the outermost poles in each row. Each coil in the drive had 200 turns. The coils were wound on a machinist lathe.

2.2. Magnetic Circuits and Magnetic Field Propagation

To create and move the vertical magnetic field patterns, circuits were sequentially energized. In Figure 1a, for example, coils B1 and B2 were energized in opposition thereby creating a North pole to form at P2. Coils T1 and T2 were opposed and with polarity opposite to the B1 and B2 coils. Therefore, P1 becomes a South pole. The magnetic field streamed from the North pole (P2) to the South pole (P1) across the air gap.

Five circuits were identified as suitable to induce and advance a strong vertical magnetic field. The mechanism considered advanced the field by one pole pitch, although the mechanism could have been extended in either direction. The circuit was numbered according to the position in sequence. The circuits were described by the energized, opposed electrical coil pair (B2, B3) and the pole direction they cause at the airgap (N or S). The circuits were:

1. (B2, B3) N, (T2, T3) S
2. (B2, B3) N, (T2, T4) S
3. (B2, B3) N, (T3, T4) S
4. (B2, B4) N, (T3, T4) S
5. (B3, B4) N, (T3, T4) S which begins the next elementary sequence.

2.3. Magnetic Measurements

We measured the device induction in the air gap as a function of coil current. The gap was 1.4 mm, which corresponded to the thickness of the MSM element used in this study. Each circuit was energized using an Uno microcontroller (Arduino, Somerville, MA, USA), which controlled an 8 channel 5 V optocoupled relay board (Sunfounder, Shenzhen, China). Ten millisecond pulses were applied and were nearly rectangular in the oscilloscope. The magnetic field was measured using a GM08 gaussmeter (Hirst Magnetic Instruments, Falmouth, UK) with a transverse Hall probe with a 1.5 mm wide Hall sensor. The measured Hall sensor location is indicated by the red box in Figure 2. For the measured circuits 1, 2, and 3, the hall probe was at locations H1, H2, and H3, respectively.

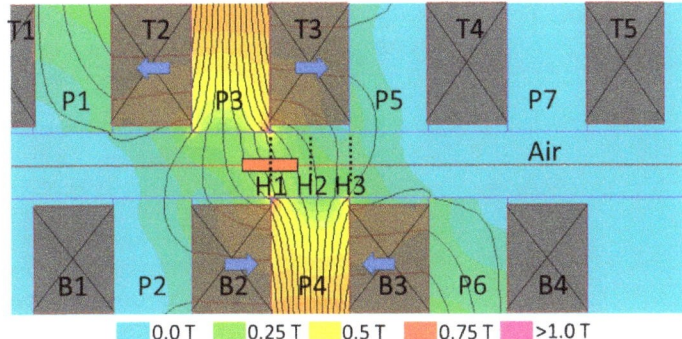

Figure 2. Schematic of the poles (P1–P7), the bottom yoke poles (B1–B4), and the top yoke coils (T1–T5). The flux density was measured in the center of the air gap, at location H1, for sequence 1. The hall probe (the red box) recorded data at positions H2 and H3 for the pulsing sequences measured experimentally.

Circuits 1, 2, and 3 were characterized by measuring the magnetic field and current at 1 V increments. The circuit was powered with a 30 V, 5 A power supply (BK Precision, Yorba Linda, CA, USA) and measured the current with a shunt resistor. An Analog Discovery 2 oscilloscope (Digilent, Pullman, WA, USA) was used to collect these data.

2.4. Actuation with MSM Element

A single crystal 10 M $Ni_{50}Mn_{28.5}Ga_{21.5}$ MSM element was used, which was manufactured at Boise State University [15]. The MSM element measured 1.4 mm × 2.0 mm × 20 mm. The faces were cut parallel to {100} crystallographic planes. The element was electropolished in a 1:3 14 M Nitric acid in ethanol solution. The element was compressed along the 20 mm dimension such that the crystallographic *c*-axes, which coincided with the direction of easy magnetization, were all aligned in the horizontal direction (Parent variant) throughout the entire element. The sample faces were mechanically ground and then finely polished using 0.3 µm diamond suspension for optical analysis. A twin domain, where the *c*-axis was aligned along the short dimension of the element, was mechanically induced into a portion of the element with calipers such that the volume fraction of the twin was approximately 15% following Ref. [12]. Then, the sample was fixed in this configuration to a glass coverslip with double-sided tape. The MSM element was actuated by circuits described in Sections 2.2 and 2.3 with an electrical current of 1 A on each coil.

The MSM element was inserted into the drive. The entire drive was placed on the stage of an optical microscope with polarized illumination. Then, circuits described in 2.2 were applied to the MSM element. Video analysis and still micrography were used to measure the twin boundary motion at the expected shrinkage location.

3. Finite Element Analysis

The change of shape and magnetization properties of the MSM material causes several effects on the MSM actuator such as a change of the inductance and magnetic permeability. Commercial finite element analysis tools are not capable of simultaneously calculating the magnetic field, mechanical stress, and resulting shape change in MSM actuators. Integrated magneto-mechanical material models are not commercially available [16,17]. Furthermore, the commonly used 10 M phase of Ni-Mn-Ga has nearly uniaxial anisotropy with a first-order uniaxial constant of $K_1(283\ K) = 2.0 \times 10^5\ J/m^3$ [18]. For example, a volume of 2.0 mm^3 twinned section in the MSM element would change the magnetocrystalline anisotropy energy of the MSM element by about 0.4 mJ.

In our model, the parent variant had a magnetic permeability of $\mu_r = 2$ in the vertical direction and 40 in the horizontal direction [19]. The twin was the opposite and had its axis of easy magnetization in the vertical direction ($\mu_r = 40$). All FEMM simulations use an energization current of 1 A/coil, producing a field in the coil in the direction indicated by arrows.

In this model, the magnetic permeability was assumed to be constant and not depend upon the field strength. Simulation of the dynamical response of the material is not straightforward. Provided a sufficiently strong magnetic field, the material switches to become twinned. Once the twin is formed, the twin becomes a low-reluctance "short" in the magnetic circuit. The position of the "short" dictates the shape of the resultant circuit. The dynamic variation of the magnetic structure was accounted for by simulating instances before and after the switching event.

3.1. Simulated Cases

The magnetic field patterns were simulated first without any MSM element for the circuits described in Section 2.2. Then, the experimental and simulated values were compared to determine the validity of the FEMM simulations.

Second, the device was coupled to an MSM element that was fully compressed, with the *c*-axis oriented horizontally. Thus, the device magnetized the MSM sample along the hard magnetization axis. The pulse sequences 1–5 were simulated as described in Section 2.2. Thus, the simulations showed the magnetic field available to cause the switching effect in an unswitched element.

Then, the magnetic field with a twin in the element was modeled. The twin had c vertical and provided a path for the flux short through the MSM element. However, we were only guessing the position of the twin. The twin was moved to different locations, determining the magnetic flux pattern and interaction of the twin with the activated poles.

Finally, the magnetic field of a permanent magnetic drive was modeled. The simulations allow for comparison of the magnetic field energy between the permanent magnetic drive and the staggered pole electromagnet.

3.2. Magnetic Field Energy

FEMM was used to analyze the magnetic field energy in the drive (with and without the MSM element). The magnetic field energy of the permanent magnet is large and varies slightly with the configuration of the MSM element. The stray field interacts with the MSM element to lower the stray field energy. FEMM integrates the magnetic field energy (MFE) of the defined region as $W_m = \int_0^B H dB$ [20].

The MFE was evaluated for two areas. The region of interest (ROI) was the region of volume occupied by the MSM element. In some simulations, the drive was simulated without the MSM element, and the material within the ROI was air. The second area was the entire system. The total system energy was approximated within the system by evaluating a cylindrical volume with 60 mm diameter and zero-flux density boundary condition on the circumference. This cylindrical volume was centered on the drive system.

Then, the MFE in the MSM element for the staggered pole circuit was considered. A 2.0 mm thick twin was introduced to the element. For each pulse sequence (1–5), the twin was moved along the element, and the MFE of the MSM element was recorded at each position. The twin was moved in increments of 0.5 mm along the element. The total system and ROI MFE energies were evaluated to make direct comparison between the energetics of the electromagnetic drive and that of the permanent magnetic drive.

4. Results

4.1. Device Measurements

Figure 3 shows the experimental results of the flux generated in the air gap by the activation of circuit 1 (gray), circuit 2 (green), and circuit 3 (blue). The drive's total current was recorded at each measurement and divided by the number of energized coils to determine the current per coil. The magnetization was quite similar for circuits 1, 2, and 3. The red curve shows the activation of only P4 by coils B2 and B3 alone. In the red curve, the magnetic induction began to saturate at a lower field in the airgap than for the other circuits, which have two energized poles.

No motion of twin boundaries was detected for the pulsing sequence. Before and after pulsing, the twin boundary geometries were identical. A slight motion appeared to be due to Maxwell forces at the poles that attracted the MSM element.

4.2. FEMM Simulation

The magnetic induction in the airgap was approximately 250 mT at 1 A on each coil, which correlated well to the experimental measurements of Figure 3.

Figure 4 shows the elementary sequence of circuits that advances the vertical flux along the single variant MSM element. In Figure 4a, circuit 1 activated poles P4 and P3. In Figure 4b, circuit 2 activated P4 and also P3 and P5. In Figure 4c, circuit 3 activated P4 and P5. In Figure 4d, circuit 4 activated P4 and P6 on the bottom yoke, and P5 on the top. The circuit number following Section 2.2 is noted in the bottom right of each frame in Figure 4.

When activated with circuit 1 (Figure 4a), poles P3 and P4 were saturated and directed magnetic flux between them. A significant portion of the flux generated by the individual poles leaked back through adjacent poles P2 and P6, rather than continuing to the other side of the MSM element. When activated with circuit 2 (Figure 4b), the vertical field in the

MSM element was broad, and some flux lines circled back to adjacent poles. Activated with circuit 3 (Figure 4c), the field pattern was symmetrical to that of circuit 1. Pole 2 leaked slightly more induction than pole 3. Activated by circuit 4 (Figure 4d), the pattern was symmetrical to circuit 2, which was advanced by half a pole pitch.

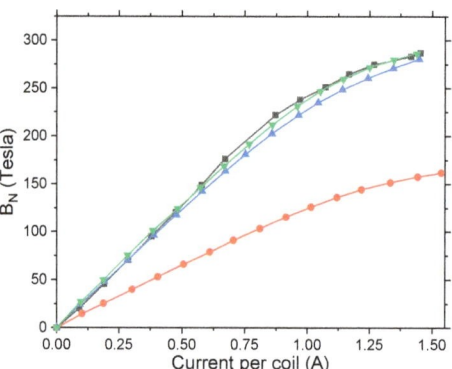

Figure 3. Experimental data of the magnetic induction in the air gap as a function of the applied current. In gray (squares), the results of pulse sequence 1, measured at Hall sensor location H1. In green (triangles pointing down), the energized sequence 2 measured at H2. The blue pulse sequence 3 (triangles pointing up), measured at H3. The red data (circles) shows activation of just the P4 pole by coils B2 and B3, at H2.

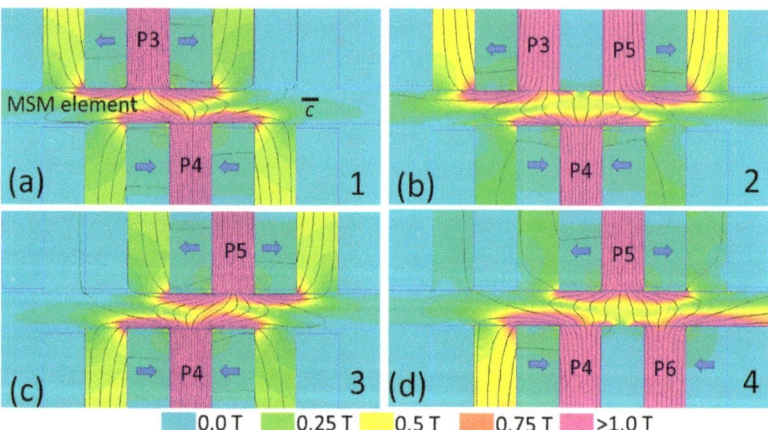

Figure 4. FEMM simulation of the pulsing sequence which approximates a moving vertical field. The simulation shows the fields induced in a single variant MSM element with the c-axis oriented horizontally. (**a**) shows circuit 1, which sent flux up and left. In (**b**) circuit 2 caused vertical and symmetrical flux. (**c**) shows circuit 3, which had identical flux to circuit 1, but mirrored in the vertical to be up and right. In (**d**), Circuit 4 had flux identical to that of circuit 2, but mirrored across horizontal and advanced a pole pitch.

Figure 5 shows the line profiles for the circuits that complete the switching pattern in the parent variant. The line position is shown in Figure 2 as the red horizontal line. The peak field takes either of two values. Circuits 1, 3, and 5 have narrower peaks and a lower peak maximum with about 0.3 T. Circuits 2 and 4 have broader peaks and a peak maximum of about 0.47 T. The broadening of the peaks is because three poles conduct the flux. The vertical field moves 6.0 mm from peak to peak, which gives the stroke of the elementary sequence.

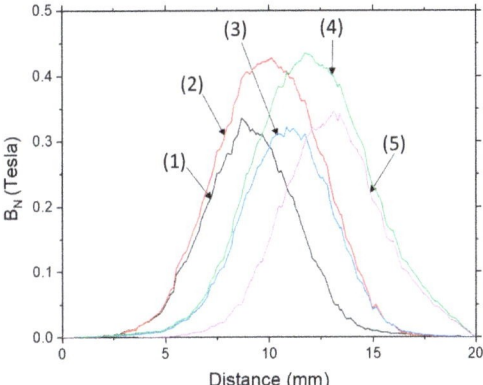

Figure 5. Profiles of the induction of the single variant element for circuits 1–5.

4.2.1. MSM Element with a Twin

The switching of the material had a large effect on induction [21]. Figure 6 shows the induction along the red line in air, the parent variant, and the parent variant with a twin, activated by circuit 1. In air, the peak of the flux was 251 mT. For the material unswitched in the parent variant, the peak maximum was 334 mT. When switched, i.e., with a twin in the MSM element, the field in the twin exceeds 1.4 T. The induction values for the twin were higher than the material's magnetic saturation of about 600 mT [16].

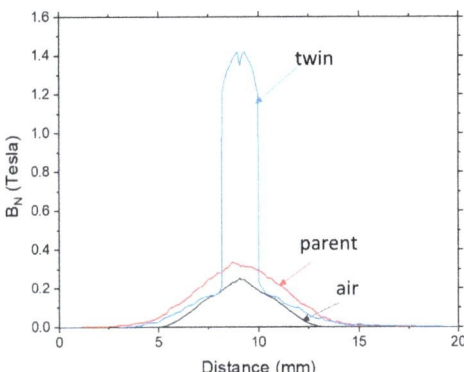

Figure 6. The twin has induction much greater than that for the parent variant, or air. The profiles show the profile along the center of the simulations presented in Figures 2, 4a and 7a.

Figure 7 shows simulation results with a twin at three different locations, namely between poles P3 and P4 (Figure 7a), between poles P4 and P5 (Figure 7b,c), and between poles P5 and P6 (Figure 7d). In the first situation, poles P3 and P4 were energized (i.e., circuit 1); in the other three cases, poles P4 and P5 were energized (i.e., circuit 3). In Figure 7a, the twin was evenly saturated across its width. The broad, symmetric peak of the twin in Figure 6 corresponds to the even saturation at the sample center, across the twin.

In Figure 7b, the magnetic flux entered the twin vertically; then, it was refracted across the twin boundary, to be horizontal in the element, before it exited the MSM element and entered P5. The right twin boundary was highly magnetized, while the left twin boundary was almost void of magnetic flux.

In Figure 7c, the flux narrowly constricted at a location defined by the connection of the right side of P4 and the left side of P5. With the advancement of the mechanism, in Figure 7d, the situation was the opposite of Figure 7b, with the left twin boundary redirecting substantial magnetic flux and the right boundary in a region of low induction. In a next step (not shown here), poles P5 and P6 were activated with the twin at the position as in Figure 7d. The resulting magnetic flux pattern was identical to that shown in Figure 7a but displaced by one pole pitch to the right.

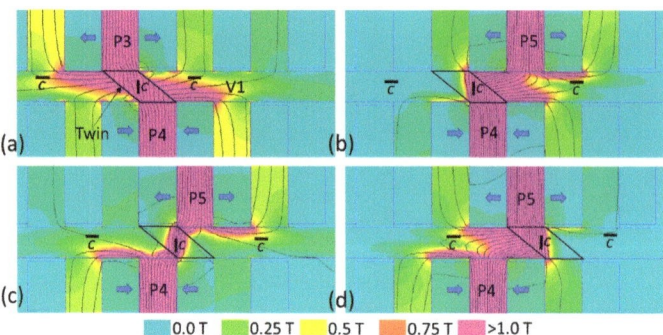

Figure 7. FEMM simulations of circuit 1 for (**a**) and circuit 3 for (**b**–**d**). The location of the twin affected the magnetic flux pattern. In (**a**), a twin variant was between P3 and P4 using circuit 1. In (**b**), with circuit 3 activated, the field interacts with the twin still at the location of (**a**). Then, the twin was moved further along the element in (**c**,**d**). In (**c**), the magnetic flux was locally concentrated in a strong vertical magnetic field. In (**d**), the magnetic flux reflected strongly across the left twin boundary, in a pattern symmetrical to that of (**b**).

Figure 8 shows profiles (a) through (d) corresponding to the simulation scenarios of Figure 7. For profile (e), the twin was in the position as in (d), but circuit 5 was energized. For this situation, the magnetic flux pattern was nearly identical to that of Figure 7a, though it was advanced by a pole pitch. The induction calculated by FEMM was higher than possible for the MSM sample, which has a saturation magnetization of about 600 mT [16]. The deviation was due to FEMM's linear approximation of the anisotropic magnetic permeability.

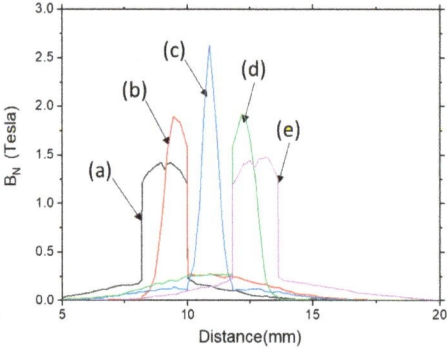

Figure 8. Simulation of the flux density along the magnetic shape memory (MSM) element, including a twin region. The twin region moved along the element corresponding to the (**a**–**d**) sequences in Figure 7. (**e**) profile shows the simulation of the (**d**) twin boundary position, using circuit 5.

4.2.2. Simulation of Magnetic Field Energies (MFE)

Figure 9 shows the MFE for each magnetic circuit (1–5) as a function of twin positions. The units of MFE were recorded in mJ/mm, which gives the energy for each mm depth of our two-dimensional (planar) simulation. Initially, the MFE for the parent variant MSM element was recorded without any twin, which is represented in Figure 9 as dashed horizontal lines with the same color as the active circuit.

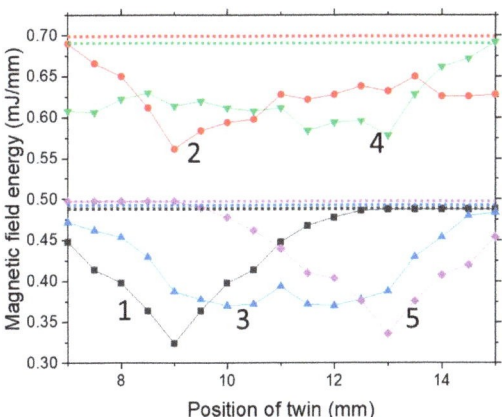

Figure 9. Drive energy at configurations of a 2.0 mm wide twin moving through the studied circuits (1–5). The dotted lines correspond to the circuit by color and give the energy of the single variant condition.

Circuits 2 and 4 had higher MFE than 1, 3, or 5. The deviation in energy between the single variant (dashed line) and the energy minima was the energy variation available to nucleate the twin. The energy reduction of circuit 2 compared to the single variant condition (dashed line) had a maximum of 0.12 mJ/mm at 9 mm. The local minima at 9 mm was distinct. The energy variation between adjacent positions was steep on the left side of the minima, yet shallow on the right side. The energy configurations for circuit 4 are nearly symmetrical to those of circuit 2.

Circuits 1, 3, and 5 had lower MFE for the single variant condition. The MFE for the three circuits was about 0.49 mJ/mm. The energy variation of circuit 1 was 0.155 mJ/mm, which is slightly greater than that of circuit 2. Circuit 1 had a clear and distinct energy minimum compared to circuit 2.

Circuit 3 had similar single variant MFE to circuits 1 and 5. The circuit does not have a distinct minimum, though. The profile takes two shallow troughs separated by a slight peak, which arose at the connection of the troughs. This behavior is interpreted as the twin being relatively stable from about 9 to 13 mm, but with little preference for position.

Figure 10 shows results of simulation of the permanent magnetic drive (PM). In Figure 10a, the magnet was surrounded by air. The system energy was found within the 60 mm boundary condition. The element energy was found within the dashed black box indicated in Figure 10a. Figure 10b shows simulation now with a parent region in the dashed box. The magnetic stray field, which previously entered the air-gap in Figure 10a, was directed horizontally to the ends of the parent variant and back down to the permanent magnet. In Figure 10c, the region was a single variant, which has the c axis oriented vertically (i.e., perpendicular to the axis of the MSM element). The magnetic anisotropy directed magnetic flux up, causing a large magnetic stray field, similar to the magnetic flux distribution in the airgap. In Figure 10d, a twin was introduced, which split the element in half. The parent variant (with c horizontal) was on the left and the twin (with c vertical) was on the right. Flux concentrates in the parent variant and only weakly magnetizes the twin. In Figure 10e, the situation was reversed. Here, the twin magnetized strongly underneath the twin boundary, and in the parent variant, it refracted across the twin boundary. In Figure 10f, a twin was created between two parent variants. Similarly, to Figure 10e, magnetic flux concentrated underneath the twin boundary. The parent variant on the left directed flux horizontally. A minimal stray field was present above the element in Figure 10b,c.

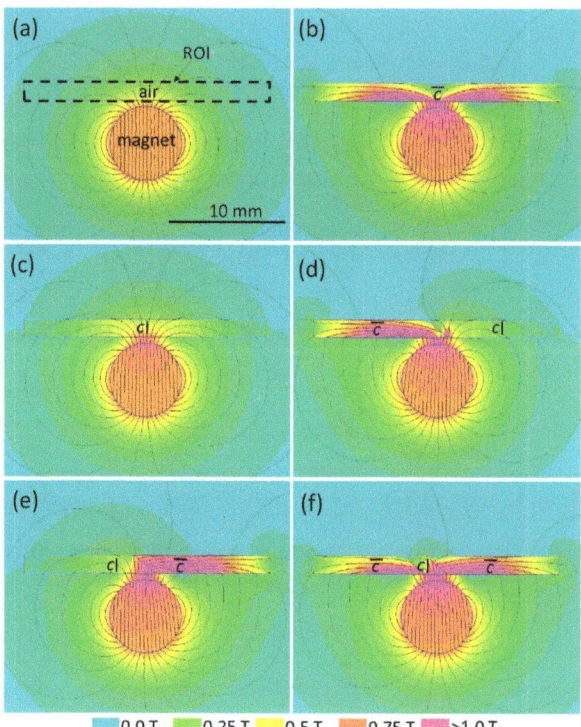

Figure 10. FEMM simulation of the magnetic flux pattern of the permanent magnetic drive. In (**a**), the cylindrical magnet generated a magnetic dipole field in air. In (**b**), the boxed region was defined as the parent variant with the c-axis oriented horizontally. (**c**) shows the boxed region defined as a single variant of twin with the c axis oriented vertically. In (**d**), a twin boundary in the center separated the twin and parent into equal volumes. The left volume was defined as the parent (c horizontal), and the right was defined as the twin (c vertical). In (**e**), a twin boundary again separates equal twin volumes; however, the twin was defined on the left and the parent was defined on the right. In (**f**), two twin boundaries were inserted with a twin in between, at the center of the parent variant. The energy within the dashed box is highest in air and lowest with two twin boundaries.

Figure 11 shows a comparison of the magnetic energies of the permanent magnetic drive and the staggered pole electromagnetic (EM) drive. The horizontal axis indicates the simulation. (a) gives the energies corresponding to Figure 10a. The black circle is the system energy. The upside-down triangles show the magnetic field energy in the magnet (gray) and the stray field (blue). The red square indicates the magnetic field energy in the boxed region, which in this simulation was air. The red squares are read by the right axis, which shows finer energy variations. Thus, for (a), the system energy was 13.62 mJ/mm, the MFE in the permanent magnet was 7.08 mJ/mm, and the MFE for the stray field was 6.54 mJ/mm. The MFE in the ROI was 0.61 mJ/mm. In (b), the air region was changed to the parent variant, and the MFE for the system and ROI were defined. The system and ROI energy are decreases relative to (a). In (c), the parent variant was changed to the twin variant, and it shows switching of the entire element. The system and ROI energy were greater than for the parent variant but less than for the air. In (c,d), a twin boundary split the volume of the ROI equally into twin and parent. With the parent on the right (d), the system energy is in-between that of the single variant parent or twin. The ROI energy is about the same as for the parent. With the parent on the left, the system energy was slightly decreased, and the ROI energy decreased more substantially. In (e), with a single thick twin in the center of two parent variants, the system energy and the ROI energy are the lowest for the permanent magnetic drive.

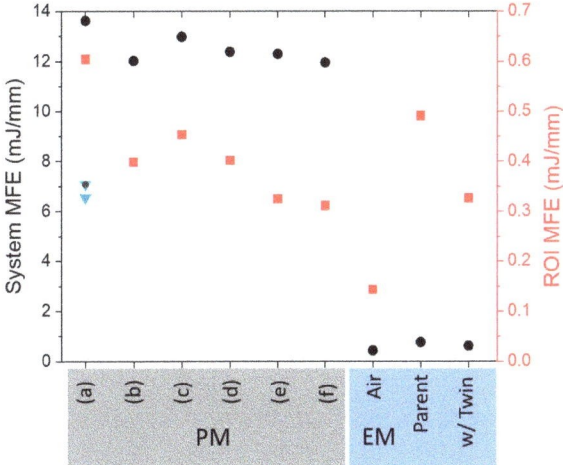

Figure 11. Comparison of the magnetic field energies for the Figure 10 configurations of the permanent magnetic (PM) to configurations of the electromagnetic (EM) drive. The black circles show the magnetic field energy of the entire system read on the left axis. Red squares show the magnetic field energy of the ROI region and reads according to the right axis. The triangles show the magnetic field energy in the magnet (black triangle) and in the magnets stray field (blue triangle) for simulation of the permanent magnet in air. (**a**) shows the energies for the permanent magnet drive in air, the magnet with single variant parent configuration (**b**) and single variant twin configuration (**c**). (**d**) with a twin boundary in the center separating parent and twin variants on the right and left, and (**e**) with the variants switched in position. (**f**) shows the magnetic field energy associated with a band of twin variant between two parent variants.

Figure 11 shows the MFE for three simulations of the electromagnetic (EM) drive for the activation of circuit 1. With an airgap separating the top and bottom yokes, the total system MFE was 0.43 mJ/mm. The MFE in the airgap region was 0.14 mJ/mm. With a parent variant MSM element inserted, the system's MFE energy increased to 0.75 mJ/mm. The MFE of the MSM element was 0.49 mJ/mm. With a twin connecting poles P4 and P3, as in Figure 7a, the systems energy was slightly lower, and the energy of the MSM element decreased compared to the parent variant, at 0.33 mJ/mm. The system MFE of the electromagnetic drive was almost an order of magnitude less than that for the permanent magnetic drive.

5. Discussion

There is a significant difference between the actuator properties of the permanent magnetic drive and the electromagnetic drive. The permanent magnetic drive creates a twin in the material and then translates the twin along the MSM element [8]. The electromagnetic drive does not readily generate the twin nor move the twin boundaries. For the permanent magnetic drive, the vertical magnetic field that first caused deformation was about 200 mT [8]. The electromagnetic drive here generates a 300 mT magnetic field in the airgap. Our previous device in [10] required a field measured in the airgap of 500 mT to nucleate a twin, which is more than three times more than the switching field of 150 mT measured along the short direction in a vibrating sample magnetometer. Thus, the local magnetic switching field is perhaps three times the global magnetic switching field.

The MFE was calculated using FEMM. The method disregards the contributions of the Zeeman energy and lowering of anisotropy energy by variant reorientation. The anisotropy energy was found in Section 3.2 to be about 0.6 mJ for a 3.0 mm^3 twin band, which is small for the permanent magnet yet significant for the EM drives. Significant sections of our element exhibit magnetic saturation, which causes additional significant error to the MFE.

In Figure 11a, the energy of the permanent magnet and the stray field without an MSM element is 13.6 mJ/mm. The permanent magnet in a PM drive is 12.7 mm long. The multiplication of these two numbers gives an MFE of 172 mJ for the system. The minimum required length of the magnet is unknown; however, if it was only as wide as the element (2.0 mm), the MFE would be 27.2 mJ. The energy needed to cause a shrinkage to form and move through the element has been experimentally measured by Smith et al. as 0.77 mJ by the difference in energy of a motor before and after coupling to the MSM element [9]. In Figure 11, by taking either the parent single variant or the twin within the parents, the stray field is reduced by about 1 mJ per mm depth of simulation. Thus, for a 2.0 mm wide MSM element, the energy reduction would be 2.0 mJ. This is perhaps not the only energy given to the MSM element, which can draw also 0.6 mJ of anisotropy energy.

For the EM drive, the MSM element can draw energy proportional to its reluctance. The maximum system energy is 0.76 mJ/mm depth, which translates for our physical device, which has a yoke depth of only 3 mm, to a maximum output of 2.28 mJ.

The variation of the stray field energy for just the MSM element is actually greater in the electromagnetic drive. A permanent magnet acts as a constant source of magnetic flux, analogous to a current source in electronics. In contrast, the electromagnetic coils act as a generators of flux potential, analogous to a voltage driving a current according to the circuit's resistance. The flux induced by the coils is a function of the reluctance of the circuit. When the reluctance is decreased, e.g., by insertion of the MSM element, the magnetic flux across the entire system increases. The different behaviors can be seen in Figure 11. When the MSM element is added to the magnetic circuit of the permanent magnetic drive, the total system energy decreases. The electromagnetic drive initially has a low system energy, as the airgap causes a large reluctance gap and reduces generated flux in the system. When the MSM element is inserted, the system energy increases by 79%.

Our device has a relatively high physical pole density. However, this does not necessarily translate to the creation of a high density of stable twin positions. In Figure 9, only circuits 1 and 5 have distinct energy minima. Circuits 2 and 4 have low energy variations, and the energy minima are at the same position as for circuits 1 and 5 (i.e., no motion of the minima). Circuit 3 exhibits a shallow depression, which is composed of two shallow troughs. The slight peak between the troughs indicates that a twin at this position is a bit unstable. The local minima observed do not account for the anisotropy energy, which together with the Zeeman energies affects the total system energy configuration, which drives the deformation. However, the calculation of all energies is quite challenging.

The energies of a single domain MSM element in the field of a permanent magnetic (Figure 10b) and the energy of an MSM element with a small twin (Figure 10f) are almost identical. The creation of the twin reduces the magnetic stray field energy by only 0.07 mJ/mm while decreasing anisotropy nearly 0.6 mJ. This indicates that other energy contributions play a significant role, such as the magnetocrystalline anisotropy energy. Further analysis of this situation is required, which is beyond the scope of this study.

6. Conclusions

An MSM element was exposed to localized magnetic fields in two ways: (1) with a permanent magnet, (2) with sets of electrical coils and yokes. The magnetic field strength at the position of the MSM element was similar for the two cases and larger than the magnetic switching field. For the permanent magnet, the magnetic field nucleated twins, while it was not so for the electromagnetic drive. This variation is attributed (1) to the high energy carried in the stray field of the permanent magnet, and (2) to the coils generating a magnetic flux potential where the actual magnetic flux depends on the reluctance of the magnetic circuit. More detailed studies are required to develop a quantitative understanding of the localized switching of MSM elements in heterogeneous magnetic fields.

Author Contributions: Conceptualization, P.M. and A.A.; methodology, software, validation, investigation, data curation, writing—original draft preparation A.A.; resources, writing—review and editing, funding acquisition, project administration, supervision P.M. All authors have read and agreed to the published version of the manuscript.

Funding: This research was funded in part by Idaho HERC 599402-1.

Acknowledgments: We gratefully acknowledge: Brent Johnston, who machined the yoke and discussed the experiments. Aaron Smith was involved in the early conception of the yoke design. Jiheon Kwon performed early simulations for a similar device, which initiated this study. We acknowledge Nadar Rafla and Paul Lindquist for their supportive discussions.

Conflicts of Interest: The authors declare no conflict of interest.

References

1. Costanza, G.; Stefano, P.; Maria, E.T. IR Thermography and Resistivity Investigations on Ni-Ti Shape Memory Alloy. *Key Eng. Mater.* **2014**, *605*, 23–26. [CrossRef]
2. Lin, Y.C.; Lin, C.F. Microstructures and magnetic properties of Fe–Ga and Fe–Ga–V ferromagnetic shape memory alloys. *IEEE Trans. Magn.* **2015**, *51*, 1–4. [CrossRef]
3. Kuchin, D.; Elvina, D.; Yurii, K.k.; Alexander, K.; Victor, K.; Alexey, M.; Vladimir, S.; Jacek, C.; Krzysztof, R.; Vladimir, K. Direct measurement of shape memory effect for Ni54Mn21Ga25, Ni50Mn41. 2In8. 8 Heusler alloys in high magnetic field. *J. Magn. Magn. Mater.* **2019**, *482*, 317–322. [CrossRef]
4. Suorsa, I.; Tellinen, J.; Ullakko, K.; Pagounis, E. Voltage generation induced by mechanical straining in magnetic shape memory materials. *J. Appl. Phys.* **2004**, *95*, 8054–8058. [CrossRef]
5. Lindquist, P.; Hobza, T.; Patrick, C.; Müllner, P. Efficiency of Energy Harvesting in Ni–Mn–Ga Shape Memory Alloys. *Shape Mem. Superelasticity* **2018**, *4*, 93–101. [CrossRef]
6. Karaman, I.; Basaran, B.; Karaca, H.E.; Karsilayan, A.I.; Chumlyakov, Y.I. Energy harvesting using martensite variant reorientation mechanism in a NiMnGa magnetic shape memory alloy. *Appl. Phys. Lett.* **2007**, *90*. [CrossRef]
7. Ullakko, K.; Wendell, L.; Smith, A.; Mullner, P.; Hampikian, G. A magnetic shape memory micropump: Contact-free, and compatible with PCR and human DNA profiling. *Smart Mater. Struct.* **2012**, *21*. [CrossRef]
8. Saren, A.; Smith, A.R.; Ullakko, K. Integratable magnetic shape memory micropump for high-pressure, precision microfluidic applications. *Microfluid. Nanofluid.* **2018**, *22*. [CrossRef]
9. Smith, A.R.; Saren, A.; Jarvinen, J.; Ullakko, K. Characterization of a high-resolution solid-state micropump that can be integrated into microfluidic systems. *Microfluid. Nanofluid.* **2015**, *18*, 1255–1263. [CrossRef]
10. Ullakko, K. Magnetically controlled shape memory alloys: A new class of actuator materials. *J. Mater. Eng. Perform.* **1996**, *5*, 405–409. [CrossRef]
11. Armstrong, A.; Karki, B.; Smith, A.; Müllner, P. Traveling Surface Undulation on a Ni-Mn-Ga Single Crystal Element. Submitted for Publication. Available online: https://core.ac.uk/download/pdf/334778582.pdf#page=127 (accessed on 23 February 2021).
12. Smith, A.; Tellinen, J.; Mullner, P.; Ullakko, K. Controlling twin variant configuration in a constrained Ni-Mn-Ga sample using local magnetic fields. *Scr. Mater.* **2014**, *77*, 68–70. [CrossRef]
13. Armstrong, A.; Finn, K.; Hobza, A.; Lindquist, P.; Rafla, N.; Mullner, P. A motionless actuation system for magnetic shape memory devices. *Smart Mater. Struct.* **2017**, *26*. [CrossRef]
14. Jiles, D. *Introduction to Magnetism and Magnetic Materials*; CRC Press: Boca Raton, FL, USA, 2015.
15. Kellis, D.; Smith, A.; Ullakko, K.; Mullner, P. Oriented single crystals of Ni-Mn-Ga with very low switching field. *J. Cryst. Growth* **2012**, *359*, 64–68. [CrossRef]
16. Schiepp, T.; Maier, M.; Pagounis, E.; Schluter, A.; Laufenberg, M. FEM-Simulation of Magnetic Shape Memory Actuators. *IEEE Trans. Magn.* **2014**, *50*. [CrossRef]
17. Gómez, E.; Roger-Folch, J.; Molina, A.; Fuentes, J.A.; Gabaldón, A.; Torres, R. Modelling of magnetic anisotropy in the finite element method. *COMPEL* **2006**, *25*, 609–615. [CrossRef]
18. Heczko, O.; Straka, L.; Lanska, N.; Ullakko, K.; Enkovaara, J. Temperature dependence of magnetic anisotropy in Ni-Mn-Ga alloys exhibiting giant field-induced strain. *J. Appl. Phys.* **2002**, *91*, 8228–8230. [CrossRef]
19. Suorsa, I.; Pagounis, E.; Ullakko, K. Position dependent inductance based on magnetic shape memory materials. *Sens. Actuators a-Phys.* **2005**, *121*, 136–141. [CrossRef]

20. Meeker, D. *Finite Element Method Magnetics—Version 4.0 User's Manual*; University of Virginia: Charlottesville, VA, USA, 2006.
21. Suorsa, I.; Pagounis, E.; Ullakko, K. Magnetization dependence on strain in the Ni-Mn-Ga magnetic shape memory material. *Appl. Phys. Lett.* **2004**, *84*, 4658–4660. [CrossRef]

Article

Full Variation of Site Substitution in Ni-Mn-Ga by Ferromagnetic Transition Metals

Vít Kopecký [1,2], Michal Rameš [1], Petr Veřtát [1,2], Ross H. Colman [3] and Oleg Heczko [1,*]

[1] Institute of Physics of the Czech Academy of Sciences, Na Slovance 1999/2, 182 21 Prague, Czech Republic; vit.kopecky@fzu.cz (V.K.); ramesm@fzu.cz (M.R.); vertat@fzu.cz (P.V.)
[2] Faculty of Nuclear Sciences and Physical Engineering, Czech Technical University in Prague, Břehová 7, 115 19 Prague, Czech Republic
[3] Faculty of Mathematics and Physics, Charles University, Ke Karlovu 5, 121 16 Prague 2, Czech Republic; ross.colman@mag.mff.cuni.cz
* Correspondence: heczko@fzu.cz

Abstract: Systematic doping by transition elements Fe, Co and Ni on each site of Ni$_2$MnGa alloy reveal that in bulk material the increase in martensitic transformation temperature is usually accompanied by the decrease in ferromagnetic Curie temperature, and vice versa. The highest martensitic transformation temperature (571 K) was found for Ni$_{50.0}$Mn$_{25.4}$(Ga$_{20.3}$Ni$_{4.3}$) with the result of a reduction in Curie temperature by 55 K. The highest Curie point (444 K) was found in alloy (Ni$_{44.9}$Co$_{5.1}$)Mn$_{25.1}$Ga$_{24.9}$; however, the transition temperature was reduced to 77 K. The dependence of transition temperature is better scaled with the N_e/a parameter (number of non-bonding electrons per atom) compared to usual e/a (valence electrons per atom). N_e/a dependence predicts a disappearance of martensitic transformation in (Ni$_{45.3}$Fe$_{5.3}$)Mn$_{23.8}$Ga$_{25.6}$, in agreement with our experiment. Although Curie temperature usually slightly decreases while the martensitic transition increases, there is no significant correlation of Curie temperature with e/a or N_e/a parameters. The doping effect of the same element is different for each compositional site. The cascade substitution is discussed and related to the experimental data.

Keywords: Ni-Mn-Ga; doping; ferromagnetism; transition metals; Heusler alloy

1. Introduction

The magnetic shape memory (MSM) effect or magnetically induced reorientation (MIR) observed mostly in Ni-Mn-Ga Heusler alloys provides up to 12% deformation induced by moderate magnetic fields of less than 1 T [1,2]. To obtain this multiferroic behaviour, the material must be ferromagnetic and exhibit martensitic transformation to a ferroelastic state. Moreover, to obtain MIR, the twin boundaries formed upon transition have to be highly mobile [3–7].

Although the MSM phenomenon is extraordinary and unique among all known metallic alloys, there are important issues which hinder its transfer from basic research toward applications. One of the main issues is a temperature limit below which the effect can be utilized. There are two limiting temperatures, the martensitic transformation temperature (T_m) and the Curie temperature (T_C), since the magnetically induced reorientation effect relies on a high magnetic anisotropy in the martensite ferromagnetic state. Thus, apart from increasing transformation temperatures, the high mobility of twin boundary and high magnetocrystalline anisotropy must be maintained.

Early studies of ternary Ni-Mn-Ga showed that T_m and T_C are in competition with each other [8–11]. Increasing the concentration of Mn at the expense of Ga (Ni$_{50}$Mn$_{25+x}$Ga$_{25-x}$) increases T_m but decreases T_C until the Curie temperature of the martensitic phase becomes smaller than T_m, at about 7 at. % extra Mn [12,13]. Similar behaviour was found for the Ni$_{50+y}$Mn$_{25-y}$Ga$_{25}$ system [10]. One way or another, the MSM effect seems to be limited to

353 K at best for the strictly ternary Ni-Mn-Ga alloy [14], which is low, considering the operating condition required by many potential applications, e.g., in the automotive industry.

Since the transition temperatures cannot be increased enough by composition variation within the ternary alloy, a large effort is devoted to increase both temperatures by adding more elements into the alloy. Dopants are usually picked from the class of transition metals, e.g., Cr, Fe, Co and Cu [15–19] or the same group of elements, as Ga is often replaced by In [20] or Sn [21–23]. However, in such kind of substitution, the fulfilment of the conditions for the MSM effect is commonly ignored.

The majority of published papers have reported the substitution of just one of the elements in the parent alloy, i.e., Ni, Mn or Ga, and thus the full potential of the multiple dopants remains hidden. An exception is the work of Soto-Para et al. [15] in which they selectively substituted Fe or Co for each original element of Ni-Mn-Ga. However, the Fe-doped alloys were based on a parent composition of $Ni_{53}Mn_{22}Ga_{25}$, and the Co-doped parent composition was $Ni_{50}Mn_{25}Ga_{25}$ (at. %). Although the off-stoichiometry of Fe doped alloys was chosen in order to improve the already high transition temperatures, the effect of Fe itself on the alloy and the compositional doping site is difficult to compare with the doped stoichiometry alloys. Recently, Armstrong et al. used a systematic doping approach, by a few atomic percent of Cu and Fe doping on Ni and Mn sites, in an attempt to increase the transformation temperatures while retaining the 10 M modulated structure [24].

In summary, despite the broadness of the existing literature, there is no simple and systematic approach to transition metal doping or alloying. Moreover, due to different preparation methods, high sensitivity to the precise composition and often not well-defined chemical analyses, it is very difficult to compare results from different authors and to obtain reasonable extrapolations and estimations. In addition, there is an apparently non-monotonous dependence of transformation temperatures on doped element concentration. A strong sensitivity to various often not well-defined parameters, and a necessity to retain specific properties to achieve MSM effects, cause large difficulties for reliable ab-initio predictions [24–26].

In this work, we selected the stoichiometric Ni_2MnGa as the initial composition to obtain a well-defined baseline and doped it with three ferromagnetic transition metals. We used the same doping amount of 5 at. %, which is a convenient amount for a direct comparison with calculations, laying between the substitution of one (3.125 at. %) and two atoms (6.25 at. %) of a $2 \times 2 \times 2$ supercell commonly used in theoretical modelling. Each original element was substituted by X = Fe, Co, Ni, resulting in a series $(Ni_{45}X_5)Mn_{25}Ga_{25}$, $Ni_{50}(Mn_{20}X_5)Ga_{25}$, $Ni_{50}Mn_{25}(Ga_{20}X_5)$ (at. %). In this way, we obtained a set of nine alloys, one of them stoichiometric Ni_2MnGa.

In order to simplify the discussion on the effect of doping, we assume that the atom replacement is ideal. Although the doping changed the electronic concentration moderately, we found dramatic changes in martensitic temperatures, while the Curie point was affected much less. Moreover, the selective doping revealed a surprisingly strong site dependence. The complete data sets for all nine alloys can serve as a reference point for advanced ab-initio calculations.

2. Materials and Methods

All the investigated alloys were prepared by the same procedure. Elements with purity of at least 99.9% were arc-melted several times, to ensure good mixing, under an overpressure argon atmosphere using a MAM-1 furnace (Edmund Bühler GmbH, Bodelshausen, Germany). The weight loss upon melting was less than 0.8%, which approximately corresponds to the extra 3% of Mn weight we added in order to balance the Mn evaporation tendency, showing that the resulting composition should be close to nominal. The resulting pellet was cut into two parts, and the larger part was annealed in a sealed glass ampoule with a partial pressure of Ar atmosphere. The heat treatment was run in two steps: First, a high temperature macroscopic homogenization treatment at 1273 K for 24 h, followed by a lower temperature ordering-improvement step at 1073 K, close to the B2 → L2$_1$ structural transition, for 24 h. After that, the samples were quenched in a water bath at room temperature. Polycrystalline samples with dimensions of about 5 mm × 3 mm × 1 mm were cut from the annealed part by a spark erosion machine and roughly polished by SiC grinding paper up to the grid of 2400.

The composition of the samples, listed in Table 1, was evaluated using an energy dispersion X-ray microfluorescence (ED-microXRF) spectrometer Eagle-III μProbe (Roentgenanalytik Systeme GmbH & Co., Taunusstein, Germany). The X-ray is emitted by a Rh tube and accelerating voltage 40 kV. The beam was focused to a spot size of about 50 μm (polycapillary focusing optics) and then detected in an 80 mm^2 Si(Li) liquid nitrogen cooled detector with a resolution of about 140 eV (at MnKα). The concentrations of the elements were evaluated at 5 spots across the sample by a semiquantitative finite elements method with one standard correction, exhibiting an error of about 0.5 at. %, which is a typical error for the method [27–29].

The precise composition evaluation is a critical issue for MSM alloys. The transition temperatures, especially the martensitic transformation, are extremely sensitive to composition, and even a small deviation from the true composition can produce a misleading error. Hence, considerable caution is necessary when comparing transformation temperatures as a function of composition from various sources.

In our notation of composition, we separate the dopant and receiving sublattice from the rest, e.g., $Ni_{49.9}Mn_{24.6}(Ga_{20.4}Co_{5.1})$. For the sake of consistency, the Ni in the stoichiometry alloy $(Ni_{45.0}Ni_{5.4})Mn_{24.6}Ga_{25.0}$ is separated, too. Throughout the paper, we mark the alloys as $Ga_{20}Co_5$ and $Ni_{45}Ni_5$, respectively. Markings of all alloys and their full compositions are shown in Table 1.

Table 1. Alloys with composition (atomic percent) given by XRF and their markings by unambiguous abbreviation; e/a stands for number of valence electrons per atom, and N_e/a counts non-bonding electrons per atom (evaluated from measured composition); structure is determined at room temperature, where A, NM and 14 M represent cubic austenite, tetragonal non-modulated and 14-layered monoclinic structures, respectively.

Identifier	Composition by XRF	e/a	N_e/a	Structure
$Ni_{45}Fe_5$	$(Ni_{45.3}Fe_{5.3})Mn_{23.8}Ga_{25.6}$	7.39	3.10	A
$Mn_{20}Fe_5$	$Ni_{49.2}(Mn_{20.5}Fe_{5.3})Ga_{25.0}$	7.53	3.18	A
$Ga_{20}Fe_5$	$Ni_{49.0}Mn_{25.5}(Ga_{20.0}Fe_{5.5})$	7.73	3.28	14 M
$Ni_{45}Co_5$	$(Ni_{44.9}Co_{5.1})Mn_{25.1}Ga_{24.9}$	7.45	3.17	A
$Mn_{20}Co_5$	$Ni_{49.9}(Mn_{20.1}Co_{5.0})Ga_{25.0}$	7.60	3.25	NM
$Ga_{20}Co_5$	$Ni_{49.9}Mn_{24.6}(Ga_{20.4}Co_{5.1})$	7.78	3.35	NM
$Ni_{45}Ni_5$	$(Ni_{45.0}Ni_{5.4})Mn_{24.6}Ga_{25.0}$	7.51	3.22	A
$Mn_{20}Ni_5$	$Ni_{50.0}(Mn_{20.5}Ni_{4.4})Ga_{25.1}$	7.63	3.29	NM
$Ga_{20}Ni_5$	$Ni_{50.0}Mn_{25.4}(Ga_{20.3}Ni_{4.3})$	7.82	3.39	NM

The XRD measurements were made using a PANalytical X'Pert PRO diffractometer (PANalytical, Almeo, Netherlands) equipped with a Co tube, in divergent and parallel beam geometry. Measured pole figures revealed the textured oligocrystalline nature of the samples. Therefore, divergent beam geometry with wide slits was used for the initial phase analysis measurements. For each sample, several scans were measured with different sample orientations suggested by the measured pole figures. This allowed us to collect a sufficient number of reflections for confident phase composition analysis.

Even though no other known phases common for Ni-Mn-Ga-based alloys than those listed in Table 2 were detected in the measured scans and pole figures, due to the measurement-complicating oligocrystalline nature of the samples and practical impossibility to cover the whole space of orientations, there might be a slight chance of the presence of lower amounts of undetected additional phases. This, however, seems to be clearly excluded by magnetic measurement. After the aforementioned measurements, selected reflections for single grains were measured in the parallel beam geometry for the precise determination of the lattice parameters which are listed in Table 2.

Table 2. Structure and lattice parameters (nm) were determined at room temperature by X-ray diffraction. XRD patterns for all compositions are shown in the Supplemental File.

Identifier	a_0	a	b	c	γ	c/a	Structure
$Ni_{45}Fe_5$	0.5827(1)	–	–	–	90	–	A
$Mn_{20}Fe_5$	0.5814(1)	–	–	–	90	–	A
$Ga_{20}Fe_5$	–	0.6200(2)	0.5764(2)	0.5506(2)	90.4(2)	0.89	14M
$Ni_{45}Co_5$	0.5822(1)	–	–	–	90	–	A
$Mn_{20}Co_5$	–	0.5535(2)	–	0.6384(2)	90	1.15	NM
$Ga_{20}Co_5$	–	0.5424(2)	–	0.6630(2)	90	1.22	NM
$Ni_{45}Ni_5$	0.5818(5)	–	–	–	90	–	A
$Mn_{20}Ni_5$	–	0.5469(4)	–	0.6515(5)	90	1.19	NM
$Ga_{20}Ni_5$	–	0.5392(2)	–	0.6701(1)	90	1.24	NM

The magnetic measurements were performed using a physical property measurement system (PPMS) with a 9 T superconducting coil (Quantum Design, Inc., San Diego, CA, USA). We used a vibrating sample measurement regime for magnetization dependence on both the external magnetic field and temperature. Magnetisation vs. field measurements confirm that magnetic saturation is clearly complete by the application of a 2 T field, and so the temperature dependence of saturated magnetization was taken at the constant field of 2 T. The saturation magnetization M_{sat} and coercive field H_C were determined from the magnetization curves, as shown in Figure 1 for two alloys. Since the martensitic and austenitic phases strongly differ in their magnetic properties, the phases can be easily distinguished from the different shape of the magnetization curves, where the greater anisotropy of the martensite phase shows a harder ferromagnetic M/H response.

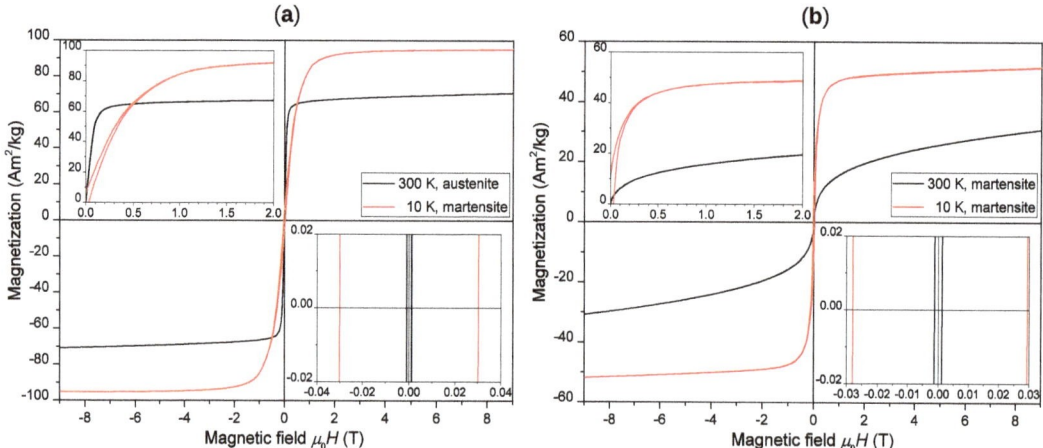

Figure 1. Isothermal magnetisation curves of (**a**) $(Ni_{45.0}Ni_{5.4})Mn_{24.6}Ga_{25.0}$ and (**b**) $Ni_{49.9}Mn_{24.6}(Ga_{20.4}Co_{5.1})$.

Owing to its high sensitivity, low-field thermomagnetic measurement ($\mu_0 H = 10$ mT) was used to determine all transition temperatures: Curie temperature T_C, premartensitic transition T_{pm}, martensite start and finish temperatures (M_s, M_f) and austenite start and finish temperatures (A_s, A_f), respectively. From the last four temperatures, the single value, equilibrium martensitic transition temperature (T_m) was calculated as $T_m = (M_s + M_f + A_s + A_f)/4$. The transition temperatures were determined from the low-field thermomagnetic curves as the intersection of two extrapolated lines, one of the maximal slope and one of the minimal slope, as shown in Figures 2 and 3. This method of transition temperature determination is common and straightforward [17,19,21,30]. The precision of the method is acceptable if the DC magnetic susceptibility or low-field magnetization measurement changes steeply in a narrow temperature range. The sharp change is common in high-quality Ni-Mn-Ga single crystals [31], and although we studied polycrystalline samples, the transition was still very steep.

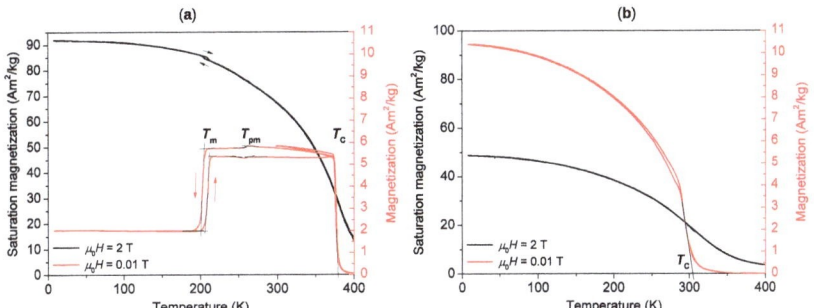

Figure 2. Low-field and saturation magnetization, measured at constant magnetic field $\mu_0 H$ of 0.01 T and 2 T, respectively, as a function of temperature: (**a**) $(Ni_{45.0}Ni_{5.4})Mn_{24.6}Ga_{25.0}$ exhibiting Curie temperature T_C, premartensitic transition T_{pm} and martensitic transformation T_m marked in the figure; (**b**) $Ni_{49.9}Mn_{24.6}(Ga_{20.4}Co_{5.1})$ exhibiting a martensitic transformation above T_C of martensite.

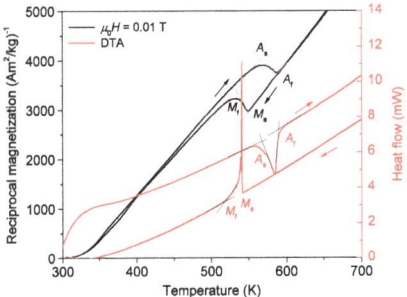

Figure 3. Reciprocal magnetization or susceptibility of $Ni_{49.9}Mn_{24.6}(Ga_{20.4}Co_{5.1})$ exhibiting martensitic transformation (MT) above Curie temperature of martensitic phase. Differential thermal analysis (DTA) measurement of the same sample confirming MT temperatures.

On the other hand, the Curie temperature determination depends on the method used. For example, the T_C of the $Ni_{45}Fe_5$ (full composition in Table 1) alloy determined by the tangent method, i.e., by slope extrapolation to zero magnetization, and by inflexion point (both of the low-field thermomagnetic curves) and by the Arrott plot [32,33] gives 399 K, 394 K and 403 K, respectively. In addition to these methods, we can determine the paramagnetic Curie temperature from a plot of reciprocal magnetization vs. T by extrapolation to zero, which gives 396 K. For consistency, we used the method of the tangents crossing for all data evaluation. Importantly, we corrected the thermomagnetic measurements for the temperature rate (4 K/min) in order to obtain the most precise transition temperatures possible. Detailed thermomagnetic curves for all samples can be found in the supplementary files.

The e/a parameter represents average number of valence electrons per atom in one formula unit. This parameter is commonly used in the field, as part of a broad common theme of valence electron counting in Heusler compounds [34]. Nevertheless, for the sake of completeness the exemplary enumeration is provided. The number of valence electrons for Ni, Mn, Ga, Fe, Co are 10, 7, 3, 8, 9, respectively, and the parameter e/a for the particular alloy with experimentally determined composition $Ni_{49.9}Mn_{24.6}(Ga_{20.4}Co_{5.1})$ is calculated as follows:

$$e/a = (10 \times 49.9 + 7 \times 24.6 + 3 \times 20.4 + 9 \times 5.1)/100 = 7.78 \tag{1}$$

Although the e/a ratio is widely used as a parameter to compare alloys of different composition, in the doped systems of Ni-Mn-Ga it becomes more complicated and anomalous behaviour appears as shown already by Ramudu et al. [35]. They introduced new parameter N_e/a and demonstrated its benefits over the e/a, e.g., the N_e/a provides the trend of T_m in the case of constant e/a for different compositions of Ni-Mn-Ga-In alloys. Another argument supporting the use the N_e/a is given later in this paper. The N_e/a parameter stands for an average number of non-bonding electrons per atom in one formula unit or so-called effective valence electrons [35]. The expression is given by the formula $N_e/a = E - N_{WS}$, where E is number of valence electrons and N_{WS} originates from the empirical model of Miedema et al. and is defined as the electron density at the boundary of the Wigner-Seitz cell [36,37]. The number N_{WS} is derived from molar volume and bulk modulus of the particular element. For the elements discussed in this work, we use non-bonding electron counts, N_{WS}, of 5.36, 4.17, 2.25, 5.55 and 5.36, for Ni, Mn, Ga, Fe and Co, respectively [35,36]. The N_e/a outcome is then enumerated similarly to e/a, for the same alloy $Ni_{49.9}Mn_{24.6}(Ga_{20.4}Co_{5.1})$:

$$N_e/a = [(10 - 5.36) \times 49.9 + (7 - 4.17) \times 24.6 + (3 - 2.25) \times 20.4 + (9 - 5.36) \times 5.1]/100 = 3.40, \tag{2}$$

Both parameters e/a and N_e/a together with the experimentally determined compositions for all prepared alloys are listed in Table 1.

3. Results

It should initially be stated that all replacements mentioned in the text strictly refer to chemical composition replacement. Additionally, we assume that the atom replacement is ideal, i.e., the substitution occurs on the atomic position of the original atom. The case of possible sequential or cascade replacement, i.e., the atoms of one element push into another sublattice, is discussed in the last section.

All doping (alloying) and comparisons are related to the very-close-to stoichiometric $(Ni_{45.0}Ni_{5.4})Mn_{24.6}Ga_{25.0}$ alloy, further on referred to as the stoichiometric alloy.

As an example, we present several figures of selected alloys in the following section to illustrate magnetic behaviour and data evaluation. The complete set of measured data and figures are provided in the supplementary files.

3.1. Transformation Temperatures and Magnetic Properties

In Figure 1, two examples of magnetization curves are displayed, $Ni_{45}Ni_5$ and $Ga_{20}Co_5$. For the stoichiometric sample, the magnetization curves clearly demonstrate the difference between magnetically soft austenite and high-anisotropy martensite. On the other hand, although the $Ga_{20}Co_5$ sample exhibits a martensitic phase at both temperatures, the magnetization curves strongly differ. The difference simply demonstrates how the magnetization curve is skewed close to the Curie point. In the austenite state and in the vicinity of the Curie point, the coercive force is negligible due to the vanishing magnetic anisotropy. From the comparison of the curves at 10 K, it follows that the Co-doped alloy exhibits a much lower saturation magnetization and also a lower magnetocrystalline anisotropy in the martensite state. The parameters determined from the magnetization curves for all prepared alloys are summarized in the Table 3.

Transformation temperatures were determined from thermomagnetic measurements; see Table 4. The examples of the measurements for the above discussed alloys are shown in Figure 2. In addition, the detection of a martensitic structural transition above the Curie point from the magnetic measurement is demonstrated in Figure 3. To confirm the transformation indicated by thermomagnetic measurements, differential thermal analysis was also performed. A reciprocal magnetization determined from thermomagnetic measurement is depicted to demonstrate its ability to detect martensitic transition (MT) in the paramagnetic state, which broadens its use to a wider range of alloys, e.g., in paramagnetic NiTi [38] and even in diamagnetic Cu-Ni-Al [39]. The measured thermomagnetic and magnetization curves for all samples are provided as supplementary information.

Table 3. Parameters determined from magnetization curves: saturation magnetization M_{sat} at 10 K and 300 K at 9 T; coercive force $\mu_0 H_C$ at 10 K and 300 K. The values corresponding to austenitic phase are marked by upper-left corner symbol $\ulcorner$.

Identifier	M_{sat}^{10K}	M_{sat}^{10K}	$\mu_0 H_C^{10K}$	M_{sat}^{300K}	$\mu_0 H_C^{300K}$
	Am²/kg	μ_B/f.u.	mT	Am²/kg	mT
Ni₄₅Fe₅	$\ulcorner$89.0	$\ulcorner$3.85	$\ulcorner$1.7	$\ulcorner$67.6	$\ulcorner$1.9
Mn₂₀Fe₅	91.4	3.96	12.2	$\ulcorner$72.0	$\ulcorner$1.8
Ga₂₀Fe₅	73.8	3.16	43.4	64.4	29.3
Ni₄₅Co₅	97.2	4.21	0.8	$\ulcorner$80.3	$\ulcorner$2.0
Mn₂₀Co₅	79.6	3.46	10.9	63.4	7.3
Ga₂₀Co₅	51.7	2.22	28.7	30.8	1.3
Ni₄₅Ni₅	95.1	4.12	30.2	$\ulcorner$70.9	$\ulcorner$1.0
Mn₂₀Ni₅	76.7	3.33	28.2	57.0	14.6
Ga₂₀Ni₅	56.9	2.44	48.5	34.1	0.9

Table 4. Transition temperatures (K) determined from thermomagnetic measurement: martensite start and finish temperatures (M_s, M_f), and austenite start and finish temperatures (A_s, A_f). From these four temperatures, the single value, equilibrium martensitic transition temperature (T_m) was calculated using formula $T_m = (M_s + M_f + A_s + A_f)/4$; Curie temperature T_C of martensite is marked *; otherwise, it is austenitic T_C, and an apparent Curie point is marked ** and discussed in Section 3.4.2; T_{pm} indicates premartensitic transition temperature.

Identifier	M_s	M_f	A_s	A_f	T_m	T_C	T_{pm}
Ni₄₅Fe₅	–	–	–	–	–	399	–
Mn₂₀Fe₅	187	170	177	195	182	411	201
Ga₂₀Fe₅	325	318	325	333	325	396	–
Ni₄₅Co₅	104	100	43	61	77	444	151
Mn₂₀Co₅	322	309	319	330	321	395	–
Ga₂₀Co₅	541	520	553	597	553	304 *	–
Ni₄₅Ni₅	205	200	206	212	206	380	258
Mn₂₀Ni₅	382	358	366	384	373	373 **	–
Ga₂₀Ni₅	564	497	603	621	571	325 *	–

Saturation magnetization, Curie and martensitic transformation temperatures, and e/a and N_e/a parameters for all samples, are summarized in Figure 4. The three main columns are plotted with respect to the doping element, and within each column, the results relate to the original element. The valence electron per atom e/a criterion is usually used for the evaluating the effect of doping. It seems that this criterion is broadly valid but not precise as pointed out before. Therefore, we used the new criterion N_e/a, as described in the Experimental Section 2, considering the amount of non-bonding electrons per atom [35,36]. In general, the martensitic transformation temperature increases broadly with increasing e/a, but the rate of increase cannot be effectively predicted. Moreover, the Curie temperature has a non-monotonous dependence similar to saturation magnetization M_{sat}. In the following, we briefly describe the differences of the magnetic behaviour affected by the substitution of individual elements Fe, Co, and Ni.

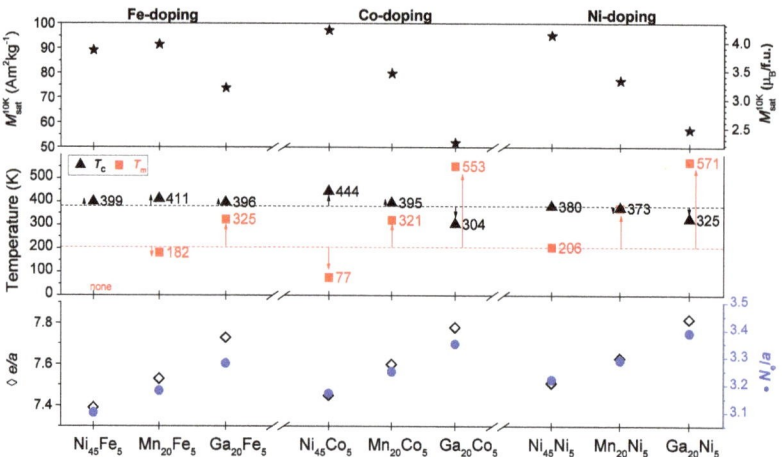

Figure 4. Transition temperatures dependence on composition. The full alloy compositions are shown in Table 1. The relative shift compared to stoichiometry alloy $Ni_{45}Ni_5$ is marked by arrows. The transformation temperatures of stoichiometry alloy are marked by dashed lines. In addition, saturation magnetization M_{sat}^{10K}, valence electrons per atom e/a and effective valence electrons per atom N_e/a parameters are shown for comparison.

3.2. Fe Substitution

3.2.1. $Ni_{45}Fe_5$

The compositional replacement of Ni with Fe increased T_C, compared to stoichiometric $Ni_{45}Ni_5$, but totally suppressed the martensitic transformation as the thermomagnetic measurement and magnetization curves down to 10 K provided no evidence of the transformation.

The austenitic structure was confirmed by X-ray diffraction at room temperature. The saturation magnetization of austenite M_{sat}^{300K} is smaller than that of the stoichiometric alloy. Coercive force H_C^{300K} (1.9 mT) is at the same level as all austenitic phases, except the stoichiometric alloy for which the H_C^{300K} is much lower (1 mT). The increase in hysteresis compared to the stoichiometric sample can be explained by a slight local deformation of the structure due to the new element and increasing disorder, which can subsequently result in a higher amount of pinning points for magnetic domain walls [40,41].

3.2.2. $Mn_{20}Fe_5$

The replacement of Mn with Fe decreases the T_m by 24 K to 182 K and increases T_C by 31 K to 411 K, compared to the stoichiometric alloy, which has the second highest Curie temperature found in this set of alloys. The $Mn_{20}Fe_5$ is the only doped alloy which exhibits a doubtless premartensitic transition, indicated by the thermomagnetic curve. The premartensitic transition was detected at a lower temperature and much closer to the martensitic transition than in the stoichiometric alloy exhibiting such transformation.

The saturation magnetization at 10 K is slightly smaller compared to the stoichiometric alloy, but it is the highest out of all Mn-deficient alloys. The H_C^{10K} = 12.2 mT in the martensite phase is less than half of that of the stoichiometric alloy. The H_C^{300K} of austenite was already discussed for $Ni_{45}Fe_5$ above, and the same is valid here. Except for H_C^{10K}, the magnetic properties are very similar to $Ni_{45}Ni_5$.

3.2.3. $Ga_{20}Fe_5$

The substitution of Ga with Fe results in a major increase in T_m by 119 K to 325 K and moderate increase of T_C by 16 K to 396 K. This alloy is one of only two alloys in the set which exhibits an increase in both T_m and T_C at the same time. Additionally, the increase is the largest observed.

Saturation magnetization at 10 K (M_{sat}^{10K}), however, is the lowest compared to the Fe-doped alloys and moderate among all alloys. The 14 M phase was expected to be preserved to 10 K as no intermartensitic transformation was detected. The value of M_{sat}^{300K} is the highest among samples in a martensite state owing to the high T_C but is still the lowest when compared to the austenites.

Overall, the Fe substitution causes an increase in T_C for all three alloys. Apparently, the introduced Fe strengthens the ferromagnetic coupling and thus stabilizes the ferromagnetic order of austenite.

3.3. Co-Substitution

3.3.1. Ni$_{45}$Co$_5$

Replacing 5 at. % of Ni with Co results in a significant change in both transition temperatures. The Curie temperature increases by 64 K to 444 K. This value of T_C is the highest of the set of alloys. Furthermore, the absolute increase in T_C from the stoichiometric alloy is almost twice as large as the Mn$_{20}$Fe$_5$, which exhibits the second highest T_C.

On the other hand, the martensitic transformation temperature T_m is very low, the lowest detected. It appears at 77 K which is 129 K below the T_m of the stoichiometric alloy. This can be compared with Ni$_{45}$Fe$_5$ in which the martensitic transformation vanished entirely.

The alloy exhibits the highest M_{sat}^{10K} and M_{sat}^{300K} of all alloys. The $H_C^{10K} = 0.8$ mT is extremely low compared to other samples in martensitic phases; it is even lower than all samples in an austenitic phase exhibited at 300 K.

3.3.2. Mn$_{20}$Co$_5$

The substitution of Mn with Co results in an increase in both transition temperatures, T_m strongly and T_C slightly compared to the stoichiometric alloy. Such coupled behaviour is rare and appears only in the Ga$_{20}$Fe$_5$ alloy. The rest of the alloys exhibit opposing trends in transition temperature. Interestingly, the transformation temperatures for Ga$_{20}$Fe$_5$ are almost identical, but the alloys differ in e/a. This demonstrates the limited validity of the e/a comparison approach. On the other hand, their N_e/a parameters differ. Compared to Ni$_{45}$Co$_5$, the coercive force H_C increased quite significantly, and it is comparable to that of Mn$_{20}$Fe$_5$.

3.3.3. Ga$_{20}$Co$_5$

Placing Co in the composition instead of Ga causes a significant increase in T_m by 347 K to 553 K, which is the second highest T_m in the set. However, the Curie temperature strongly decreases by 76 K to 304 K, and appears in the martensitic phase (T_C^M). The M_{sat}^{10K} is the lowest and nearly half that of Ni$_{45}$Co$_5$. The coercive force H_C^{10K} is at a maximum, while H_C^{300K} is very low for martensite, which is due to the vicinity of the Curie point.

3.4. Ni Substitution

3.4.1. Ni$_{45}$Ni$_5$

The substitution of Ni for Ni obviously results to the stoichiometric alloy Ni$_{50.4}$Mn$_{24.6}$Ga$_{25.0}$, or in consistent marking (Ni$_{45.0}$Ni$_{5.4}$)Mn$_{24.6}$Ga$_{25.0}$ (at. %). The measured composition is very close to stoichiometry, and the transition temperatures are in agreement with the generally accepted T_m and T_C [42,43]. Saturation magnetization at 10 K of this alloy is the second highest in the set as the stoichiometric alloy is purely ferromagnetic [44]. The coercive field of austenite at 300 K is almost half when compared to the doped alloys in the austenitic phase. Increased coercivity in some doped alloys suggests that the doping elements may locally deform the cubic lattice providing more pinning points which hinder the magnetic domain wall motion resulting in a higher coercive force than the dopant-free stoichiometric alloy.

3.4.2. $Mn_{20}Ni_5$

Placing additional Ni instead of Mn causes a significant increase in T_m by 167 K to 373 K, close to the Curie point. In this alloy, it is apparent that the Curie point and MT temperature coincide. The coalescence of T_m and T_C has been found in various compositions of Ni-Mn-Ga [10] or Ni-Mn-In- and Ni-Mn-Sn-based alloys [45,46], often coined as a metamagnetic transition. Entel et al. ascribed the effect to large volume magnetostriction in the vicinity of magnetic phase transition [10]. From our perspective, the phenomenon is simple. If the Curie temperature of martensite T_C^M is above T_m and the Curie temperature of austenite T_C lies below T_m [47], then during cooling, when crossing the martensitic transition, the paramagnetic austenite transforms directly to the ferromagnetic martensite, which has a higher Curie point. This causes the appearance of magnetic order simultaneously with the martensitic transition and apparent Curie point. During heating, the ferromagnetic martensite transforms upon MT directly to paramagnetic austenite, resulting in hysteresis of the Curie point. Thereby, the MT creates an apparent Curie temperature as the paramagnetic austenite undergoes the martensitic transformation to ferromagnetic martensite.

3.4.3. $Ga_{20}Ni_5$

The substitution of Ga with Ni increases T_m dramatically by 365 K up to 571 K, which is the highest martensitic transformation temperature detected in the set. The Curie temperature decreases again and is exhibited in martensite phase (T_C^M) at 325 K, which is 55 K below the T_C^{stoi}. This makes it the second lowest Curie point in the set. In comparison with the $Ga_{20}Co_5$ alloy, this alloy exhibits a similar temperature decrease, i.e., both Ni and Co exhibit similar effect when substituted for Ga. The M_{sat}^{10K} is the second lowest and almost half that of the highest value, which is $Ni_{45}Co_5$, and the stoichiometric alloy. The coercive force at 10 K is the highest and at 300 K the lowest of all alloys. The value at 300 K is so low due to the proximity of T_C^M and low easy plane anisotropy of NM martensite. On the contrary, $H_C^{10K} = 48.5$ mT is unusually high for NM martensite, and the closest one is $Ga_{20}Fe_5$, which is expected to be 14 M martensite at 10 K.

4. Discussion

Thanks to consistent doping by transition elements in all positions, we can test the predictive power of the widely used e/a and recently introduced N_e/a criteria for transition temperatures. Figure 5 shows the absolute values of transition temperatures T_m, T_C and T_{pm} with respect to (a) e/a and (b) N_e/a parameters. A better linear fit, i.e., statistically more significant, of T_m, is provided by the N_e/a dependency, and the premartensitic transformation temperature is also well correlated, although we have only three points. Despite the error in the chemical analysis, about 0.5 at. %, the martensitic transition temperatures of all alloys fall into a single line, $T_m = 2130 \times N_e/a - 6739$, as shown in Figure 6. Moreover, a linear fit allows the estimation of the possible T_m of the ($Ni_{45}Fe_5$) alloy. The extrapolation based on the e/a parameter gives $T_m = 1222 \times e/a - 8996 = 31.5$ K, while using the N_e/a parameter suggests that there is no martensitic transformation (Figure 6), in agreement with our experimental result. From this perspective, the N_e/a parameter has better predictive power than the e/a ratio. On the other hand, the T_C is scattered and does not exhibit a clear correlation with either parameter.

Although one can expect that the saturation magnetization M_{sat} depends mostly on Mn content, the experiment shows different results. Despite the complex behaviour depending on specific site doping, the general trend is that saturation magnetization decreases with increasing electronic concentration e/a, exhibiting an approximate maximum for stoichiometric alloy, i.e., in an ideal $L2_1$ ordered structure. The dependence is shown in the supplementary information. Moreover, the e/a parameter leads to a slightly better linear fit than N_e/a.

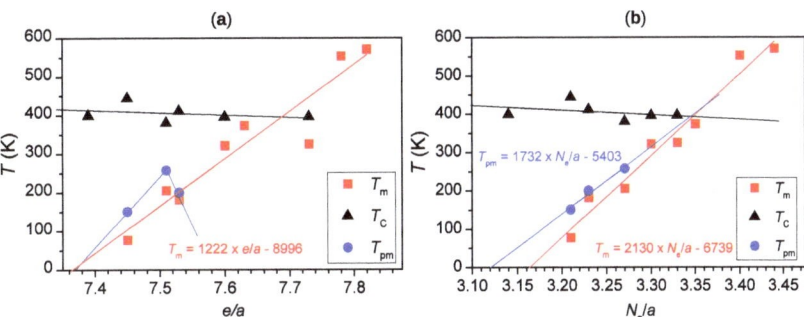

Figure 5. Premartensitic (T_{pm}), martensitic (T_m) and Curie (T_C) transition temperatures as (**a**) a function of number of valence electrons per atom (e/a) and (**b**) a function of number of non-bonding electrons per atom (N_e/a). Linear fit of T_m is well correlated to both e/a and N_e/a; T_C does not exhibit a good linear fit.

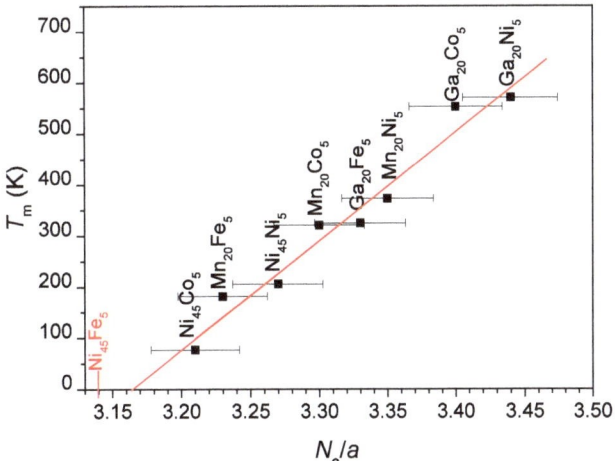

Figure 6. Martensitic transformation temperatures as a function of non-bonding number of electrons per atom N_e/a. The marked vertical line corresponds to the $(Ni_{45.3}Fe_{5.3})Mn_{23.8}Ga_{25.6}$ alloy exhibiting no martensitic transition with $N_e/a = 3.1$. Linear fit of the data provides a formula $T_m = 2130 \times N_e/a - 6739$ intersecting the value 3.1 below 0 K.

Most of the experimental studies of doped Ni-Mn-Ga have selectively substituted just one of the original elements, e.g., [16,19,48–50]. At first approximation, it is a valid approach to introduce a new element in order to study its effect on transition temperatures. However, as we have shown, it is not sufficient. The dopant placing in the composition is crucial and significantly affects the transition temperatures, which is discussed further in the text.

For Fe-doped alloys, Soto-Para et al. [15] prepared a similar full doping series substituting all original elements. However, the Fe-doped Ni-Mn-Ga series were based on the off-stoichiometric parent $Ni_{53}Mn_{22}Ga_{25}$ alloy, with an Fe content from 0 to 5 at. %. Since the base alloy exhibits significantly different transition temperatures from the stoichiometric one, the direct comparison with this work is difficult; however, the general trends are consistent. In agreement with their reasoning [15], the effect of Fe doping is to enhance magnetic exchange coupling and thus increase T_C. The doping by Fe, on the other hand, apparently caused a decrease in the free energy difference between the austenite and martensite phases, resulting in the stabilization of the austenitic parent phase, i.e., decreasing the T_m

Similarly, the Co-doped series from the same paper [15] is based on the stoichiometric alloy as the reference system, which enables a direct comparison, and the trends again agree very well. Although the absolute values are not provided but estimated, the Curie temperatures T_C seem to be in better agreement than the martensitic temperatures T_m. This may not be surprising as T_m is much more sensitive to composition. Moreover, Soto-Para et al. used the martensite start temperatures (M_s), while our approach utilized the equilibrium martensitic temperature T_m. When making similar comparisons using M_s (supplementary information), our trends are clearly comparable within, a discrepancy which can be confidently ascribed to composition variation between our alloys and the composition determination error.

Until now, we discussed how the new elements affect the transformation temperatures T_m and T_C, as summarized in Figure 4. Additionally, we showed that if the dopant is placed into various compositional sites, the transition temperatures follow broadly the same trend, regardless of which new element is introduced. This can provide an alternative perspective related to the reduction in the original element contrasting the introduction of a new one.

Following this idea, we rearranged the obtained results with respect to the deficiency of the original elements; see Figure 7. This immediately shows that if Ni is deficient, T_m decreases strongly and T_C increases significantly, irrespective of the substituting element. This trend was also observed in [16,48,51] for Co or in [52] for Fe.

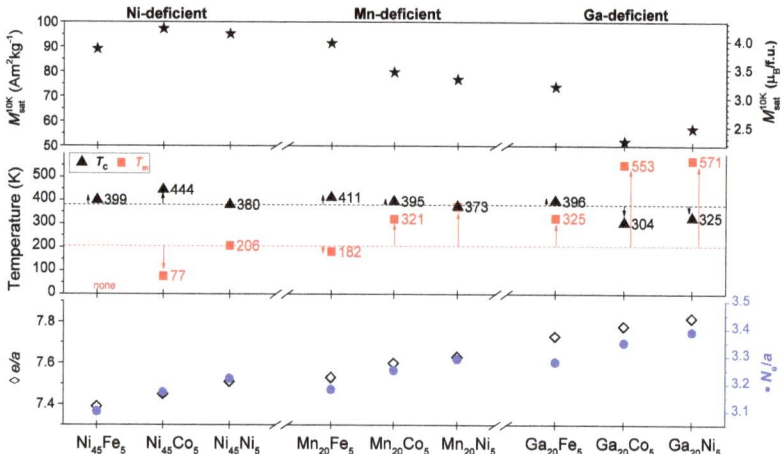

Figure 7. Transition temperatures dependences on composition rearranged according to a doping site. In addition, saturation magnetization M_{sat}^{10K} and e/a and N_e/a parameters are plotted for comparison. The dashed lines mark transition temperatures of the stoichiometric alloy, and the arrows emphasize a temperature deviation from them.

On the other hand, if Ga is deficient, the effect is almost opposite and very steep, T_m increased and T_C decreased, except for $Ga_{20}Fe_5$, where the T_C also slightly increased. Such trends are not unique and can be found in other works [15,30,53,54]. Surprisingly, the substitution of Ga by ferromagnetic metals results in a strong decrease in magnetic moment, possibly due to antiferromagnetic coupling with the major magnetic moment on Mn atoms.

The Mn-deficient alloys do not exhibit a clear trend, two out of three alloys show an increase in T_C and a different two show an increase in T_m. In ternary Ni-Mn-Ga, the deficiency of Mn leads to an opposite effectsdepending on whether the composition is Ni- or Ga-rich. The $Mn_{20}Fe_5$ exhibits a similar trend of transition shift to Ga-rich alloys [55,56] and $Mn_{20}Co_5$ and $Mn_{20}Ni_5$ to Ni-rich alloys [10,57]. Although the effect on Mn-deficient alloys is not monotonous and depends on the doping element, our data are in agreement with previous reports, e.g., $Mn_{20}Fe_5$ with [58], $Mn_{20}Co_5$ with [18,51,59] and $Mn_{20}Ni_5$ with [10,57]. It seems that the effects of the doping element is minor compared to the Mn-deficiency itself.

In reality, substitutions may not be ideal, as assumed, and the dopant may prefer another sublattice than the intentionally vacant one. Then, the dopant may push the original atoms from their own preferred lattice site into the vacant one, and this can start a cascade of substitutions between the particular sublattices. The cascade means that, e.g., Co is supposed to compositionally replace Ga written as a $Ni_{50}Mn_{25}(Ga_{20}Co_5)$ alloy, but in fact, the Co atoms tend to occupy, for instance, the Mn sublattice, and the Mn atoms are pushed into the vacant Ga sublattice forming the $Ni_{50}(Mn_{20}Co_5)(Ga_{20}Mn_5)$ alloy. We call this type a cascade of a first order.

Of course, if we extend the idea one step further from the first-order cascade, it results into the second-order cascade. If in the previous example Co would prefer the Ni sublattice, the expelled Ni would prefer the Mn sites, and Mn must settle in the vacant Ga sublattice; the composition formula could be then written according to the preferential sites as $(Ni_{45}Co_5)(Mn_{20}Ni_5)(Ga_{20}Mn_5)$.

It is difficult to determine preferential sites experimentally. The similarities in X-ray scattering factors result in very little contrast between these elements with such close electron numbers. Neutron diffraction can provide an enhanced contrast, but the possibility of refining the occupancy of three elements on a single site requires the careful use of either multiple data sets from different radiation types, or careful imposed constraints that must be reasonably justified.

Richard et al. studied four Ni-Mn-Ga compositions from which three were powdered single crystals and one was maintained for single crystal diffraction [60]. The best model fitting the neutron diffraction data suggested that excess Mn atoms tend to occupy Ni and Ga sites and no cascade occurs. On the other hand, for the Ni-rich alloy, the atoms expel the Mn into the Ga sites, i.e., the first-order cascade, resulting in enhanced ferromagnetic (FM) coupling on the Mn sites, and enhanced antiferromagnetic (AFM) coupling on the Ga sites.

For the doped Ni-Mn-Ga alloy, there are three neutron diffraction reports which deal with the site preference. In the report by Porro-Azpiazu et al., they studied the $Ni_{51}(Mn_{28-x}Y_x)Ga_{21}$ (Y= Fe, Co) system but other than the available experimental report, the results have not been published as a peer reviewed article [61]. A more complicated system of $(Ni_{45}Co_5)(Mn_{25-x}Fe_x)(Ga_{20}Cu_5)$ ($x = 0, 1, 2, \ldots, 6$ and 8) was studied by Lazpita et al. mostly on powder samples prepared from single crystals [62]. Although their intention was obviously to replace the original elements by the particular ones as marked, the dopants were found in different sites (for $x = 1$), resulting in complex situation: Cu in Mn sites, which expels Mn into Ga and Ni sites; Co in Ga sites; Fe in Ni and Mn sites. The following report from two single crystals of compositions $x = 4$ and 5 confirmed the aforementioned results [63]. However, these reports have also not yet been published, apparently due to the complex nature of the substitution, and the required complexity of the analysis. Although one can expect that different elements favour specific sublattices, firm evidence is still missing.

Since the site preference is difficult to study experimentally, and an insufficient amount of data has so far been provided, we resort to reported ab-initio calculations [64,65]. Li et al. [64] showed that on a system doped by 1.25 at. % of Fe, Co or Cu, the energetically preferable sites for Fe are Ni and Mn sublattices when they are vacant. Rather than to settle in a Ga vacant sublattice, Fe causes a cascade with Mn expelled into the Ga sublattice. However, the calculated energy of the cascade is only slightly more preferable than that of the cascade with Ni or the no cascade system. As the difference between calculated energies is not conclusive at all, it means one should be able to find Fe in Ga sites at least up to 1.25 at. %. This correlates with the neutron diffraction report by Perez-Checa et al. [63] in which they found 0.5 at. % of Fe in the Ga sublattice out of 5.25 at. % doping.

It may be that only a small amount of Fe can be incorporated into the Ga sublattice, and the rest settle in Mn and Ni sites. However, their studied system, having three different dopants, is complex, and the reported settlement may be a product of this complexity or disorder.

The results presented here indicate the first-order cascade with Mn sites and the combined effect of both Fe and Mn on their doping sites. Fe in the Mn sublattice results in a minor decrease in T_m and a major increase in T_C, as shown in Figure 4, and may substitute Mn adequately, as they have similar atomic radii. Expelled Mn in the Ga sublattice causes a steep increase in T_m and a slight decrease in T_C, which is widely known from the Mn-rich Ni_2MnGa. Both effects combined may result in transition temperatures found in the $Ga_{20}Fe_5$ alloy.

For the Co doped system, Li et al. [64] showed by ab-initio calculations that Co atoms tend to occupy the Ni sublattice in every scenario and push Ni atoms into the vacant sublattice. The second-order cascade, i.e., Co expelling Ni and Ni expelling Mn, is not energetically preferred as much as the first-order cascade, according to Li et al. This partially disagrees with the experimental data of Richard et al. [60] and calculations of Jie et al. [65], both conducted on a Ni-rich Ni-Mn-Ga alloy, showing that extra Ni prefers Mn sites. Therefore, the second-order cascade could appear in the Co-doped Ga-deficient alloys.

Moreover, a relatively simple quantitative model was proposed by Ayila et al. [66]. They studied the occupancy of Co and Fe in a Ga-deficient Ni-Mn-Ga alloy by the calculation of magnetic moment in several occupancy and magnetic order configurations. They showed that both Fe and Co prefer to settle in the filled Mn sublattice, and to align their magnetic moment parallel to the overall ferromagnetic moment of Mn and Ni atoms. The displaced Mn occupies the vacant Ga sites with AFM aligned magnetic moment. Both Fe and Co atoms begin the cascade substitution which is for Fe in agreement with ab-initio calculations [64] and neutron diffraction report [63] but in disagreement for Co. Although the proposed mechanism is in agreement with our experiment, i.e., the saturation magnetization of alloys doped on Ga sites are the lowest, it is inconclusive to ascribe the observed sharp decrease to only Mn antiferromagnetically ordered in Ga position.

When we apply the model of Ayila et al. [66] to our data, it suggests that at least the first-order cascade takes place for the Ga-deficient alloys. Doping the Ni-deficient alloys indicates that Fe and Co tend to settle on the Ni sites being FM coupled. The predicted values for Mn-deficient alloys differ for each case. Fe may settle directly in the Mn sublattice and couple FM, Ni may also settle directly on the Mn sites but couple AFM, which disagrees with the neutron diffraction experiment [60]. In the case of Co, the best correlation with experimental values gives the first-order cascade when Co (FM) pushes Ni into the Mn sublattice, coupled AFM again. This kind of modelling, though, is critical in tuning the parameters close to the experimental values. In order to receive a better fit for some alloys, the elements would need to be mixed much more, and some sort of collective moment should be added.

5. Conclusions

By consistent and selective doping using transition elements (5 at. %) to stoichiometric Ni_2MnGa, the highest Curie temperature $T_C = 444$ K was achieved in the $(Ni_{44.9}Co_{5.1})Mn_{25.1}Ga_{24.9}$ alloy. This alloy, however, exhibited the lowest martensitic transition temperature $T_m = 77$ K detected. Similarly, doping by Fe in the alloy $(Ni_{45.3}Fe_{5.3})Mn_{23.8}Ga_{25.6}$ stabilized the austenitic phase down to 10 K.

The highest temperature of martensitic transformation $T_m = 571$ K was achieved for the $Ni_{50.0}Mn_{25.4}(Ga_{20.3}Ni_{4.3})$ alloy. However, this doping also resulted in a significant reduction in Curie temperature to 325 K.

From the obtained experimental data, it is clear that to broaden the interval, where the MSM effect occurs, is impossible by single element doping within this substitutional range.

The well-established parameter e/a was compared to the N_e/a parameter. Martensitic transformation temperatures increase with increasing e/a and N_e/a parameters, with the latter providing a better linear fit and predicting the disappearance of martensitic transformation in the Fe-doped alloy.

The Curie temperature consistently increases in Fe-doped alloys and decreases in Ni-doped alloys. Co-doped alloys exhibit either an increase or a decrease in T_C depending on the element that is substituted for. There is no significant correlation between Curie temperature and e/a or N_e/a parameters.

The doping effect is different for each compositional site; therefore, it is necessary to substitute for all three positions to reveal the full potential of the dopant. We suggest that the deficiency of the original element may affect the transition temperatures more strongly than the incorporation of a new element itself, at least up to 5 at. %. Although the site preference of the dopant is difficult to obtain experimentally, the cascade substitution is discussed and related to experimental data.

Apart from the valuable insight into doping by transitional elements, we expect that this experimental work can be treated as an incentive for ab-initio calculation explaining our observed trends.

Supplementary Materials: The following are available online at https://www.mdpi.com/article/10.3390/met11060850/s1; Figures S1–S9: magnetization and thermomagnetic curves of all alloys; Figures S10 and S11: Curie and martensitic transformation temperatures with respect to alloy composition; Figure S12: Martensite transformation start temperature with respect to e/a or N_e/a parameters; Figure S13: Saturation magnetization with respect to e/a or N_e/a parameters; Figure S14: Summary of X-ray diffraction patterns for all alloys.

Author Contributions: Conceptualization, V.K. and O.H.; validation, V.K. and O.H.; formal analysis, V.K. and P.V.; investigation, M.R., R.H.C. and P.V.; writing—original draft preparation, V.K.; writing—review and editing, V.K., O.H., R.H.C., P.V. and M.R.; supervision, O.H.; funding acquisition, O.H. All authors have read and agreed to the published version of the manuscript.

Funding: This work was supported by Czech Science Foundation, grant No. 19-09882S, and Czech OP VVV projects SOLID21-CZ.02.1.01/0.0/0.0/16_019/0000760 and MATFUN—CZ.02.1.01/0.0/0.0/15_003/0000487. Magnetic measurements were performed in MGML http://mgml.eu (accessed on 14 April 2021), which is supported within the program of Czech Research Infrastructures (project no. LM2018096). V.K. and P.V. would like to acknowledge financial support from the grant SGS19/190/OHK4/3T/14.

Institutional Review Board Statement: Not applicable.

Informed Consent Statement: Not applicable.

Data Availability Statement: The data presented in this study are available in the Supplementary Material.

Acknowledgments: Authors wish to thank Tomáš Kmječ for the XRF analysis.

Conflicts of Interest: The authors declare no conflict of interest. The funders had no role in the design of the study; in the collection, analyses, or interpretation of data; in the writing of the manuscript, or in the decision to publish the results.

Abbreviations

The following abbreviations are used in this manuscript:

MSM	magnetic shape memory
MT	martensitic transformation
FM	ferromagnetic
AFM	antiferromagnetic

References

1. Sozinov, A.; Lanska, N.; Soroka, A.; Zou, W. 12% magnetic field-induced strain in Ni-Mn-Ga-based non-modulated martensite. *Appl. Phys. Lett.* **2013**, *102*, 21902. [CrossRef]
2. Heczko, O.; Sozinov, A.; Ullakko, K. Giant field-induced reversible strain in magnetic shape memory NiMnGa alloy. *IEEE Trans. Magn.* **2000**, *36*, 3266–3268. [CrossRef]
3. Heczko, O.; Scheerbaum, N.; Gutfleisch, O. Magnetic shape memory phenomena. In *Nanoscale Magnetic Materials and Applications*; Liu, J.P., Fullerton, E., Gutfleisch, O., Sellmyer, D.J., Eds.; Springer: New York, NY, USA, 2009; Chapter 14, pp. 399–439. [CrossRef]
4. Sozinov, A.; Lanska, N.; Soroka, A.; Straka, L. Highly mobile type II twin boundary in Ni-Mn-Ga five-layered martensite. *Appl. Phys. Lett.* **2011**, *99*, 124103. [CrossRef]
5. Straka, L.; Soroka, A.; Seiner, H.; Hänninen, H.; Sozinov, A. Temperature dependence of twinning stress of Type I and Type II twins in 10 M modulated Ni-Mn-Ga martensite. *Scr. Mater.* **2012**, *67*, 25–28. [CrossRef]
6. Straka, L.; Hänninen, H.; Soroka, A.; Sozinov, A. Ni-Mn-Ga single crystals with very low twinning stress. *J. Phys. Conf. Ser.* **2011**, *303*, 12079. [CrossRef]
7. Kopecký, V.; Perevertov, O.; Straka, L.; Ševčík, M.; Heczko, O. Equivalence of mechanical and magnetic force in magnetic shape memory effect. *Acta Phys. Pol. A* **2015**, *128*, 754–757. [CrossRef]
8. Lanska, N.; Söderberg, O.; Sozinov, A.; Ge, Y.; Ullakko, K.; Lindroos, V.K. Composition and temperature dependence of the crystal structure of Ni-Mn-Ga alloys. *J. Appl. Phys.* **2004**, *95*, 8074–8078. [CrossRef]
9. Castán, T.; Vives, E.; Mañosa, L.; Planes, A.; Saxena, A. Disorder in Magnetic and Structural Transitions: Pretransitional Phenomena and Kinetics. In *Magnetism and Structure in Functional Materials*, 1st ed.; Planes, A., Mañosa, L., Saxena, A., Eds.; Springer: Berlin/Heidelberg, Germany, 2005; Volume 79, Chapter 3, pp. 27–48. ISBN 978-3-540-23672-6. [CrossRef]
10. Entel, P.; Buchelnikov, V.D.; Khovailo, V.V.; Zayak, A.T.; Adeagbo, W.A.; Gruner, M.E.; Herper, H.C.; Wassermann, E.F. Modelling the phase diagram of magnetic shape memory Heusler alloys. *J. Phys. D Appl. Phys.* **2006**, *39*, 865. [CrossRef]
11. Vasil'ev, A.N.; Buchel'nikov, V.D.; Takagi, T.; Khovailo, V.V.; Estrin, E.I. Shape memory ferromagnets. *Phys-Usp+* **2003**, *46*, 559. [CrossRef]
12. Xu, X.; Nagasako, M.; Ito, W.; Umetsu, R.Y.; Kanomata, T.; Kainuma, R. Magnetic properties and phase diagram of $Ni_{50}Mn_{50-x}Ga_x$ ferromagnetic shape memory alloys. *Acta Mater.* **2013**, *61*, 6712–6723. [CrossRef]
13. Jin, X.; Marioni, M.; Bono, D.; Allen, S.C.; O'Handley, R.; Hsu, T.Y. Empirical Mapping of Ni-Mn-Ga Properties with Composition and Valence Electron Concentration. *J. Appl. Phys.* **2002**, *91*, 8222–8224. [CrossRef]
14. Pagounis, E.; Chulist, R.; Szczerba, M.; Laufenberg, M. High-temperature magnetic shape memory actuation in a Ni–Mn–Ga single crystal. *Scr. Mater.* **2014**, *83*, 29–32. [CrossRef]
15. Soto-Parra, D.; Moya, X.; Mañosa, L.; Planes, A.; Flores-Zúñiga, H.; Alvarado-Hernández, F.; Ochoa-Gamboa, R.; Matutes-Aquino, J.; Ríos-Jara, D. Fe and Co selective substitution in Ni2MnGa: Effect of magnetism on relative phase stability. *Philos. Mag.* **2010**, *90*, 2771–2792. [CrossRef]
16. Kanomata, T.; Kitsunai, Y.; Sano, K.; Furutani, Y.; Nishihara, H.; Umetsu, R.Y.; Kainuma, R.; Miura, Y.; Shirai, M. Magnetic properties of quaternary Heusler alloys $Ni_{2-x}Co_xMnGa$. *Phys. Rev. B* **2009**, *80*, 214402. [CrossRef]
17. Sakon, T.; Fujimoto, N.; Kanomata, T.; Adachi, Y. Magnetostriction of $Ni_2Mn_{1-x}Cr_xGa$ Heusler Alloys. *Metals* **2017**, *7*, 410. [CrossRef]
18. Gomes, A.M.; Khan, M.; Stadler, S.; Ali, N.; Dubenko, I.; Takeuchi, A.Y.; Guimarães, A.P. Magnetocaloric properties of the $Ni_2Mn_{1-x}(Cu,Co)_xGa$ Heusler alloys. *J. Appl. Phys.* **2006**, *99*, 08Q106. [CrossRef]
19. Adachi, Y.; Kouta, R.; Fujio, M.; Kanomata, T.; Umetsu, R.Y.; Xu, X.; Kainuma, R. Magnetic Phase Diagram of Heusler Alloy System $Ni_2Mn_{1-x}Cr_xGa$. *Phys. Proc.* **2015**, *75*, 1187–1191. [CrossRef]
20. Xu, X.; Yoshida, Y.; Omori, T.; Kanomata, T.; Kainuma, R. Magnetic Properties and Phase Diagram of $Ni50Mn_{50-x}Ga_{x/2}In_{x/2}$ Magnetic Shape Memory Alloys. *Shape Mem. Superelasticity* **2016**, *2*, 371–379. [CrossRef]
21. Kanomata, T.; Shirakawa, K.; Kaneko, T. Effect of hydrostatic pressure on the Curie temperature of the Heusler alloys Ni_2MnZ (Z = Al, Ga, In, Sn and Sb). *J. Magn. Magn. Mater.* **1987**, *65*, 76–82. [CrossRef]
22. Chatterjee, S.; Giri, S.; De, S.; Majumdar, S. Giant magneto-caloric effect near room temperature in Ni–Mn–Sn–Ga alloys. *J. Alloys Compd.* **2010**, *503*, 273–276. [CrossRef]
23. Chatterjee, S.; Giri, S.; Majumdar, S.; De, S.; Koledov, V. Effect of Sn doping on the martensitic and premartensitic transitions in Ni_2MnGa. *J. Magn. Magn. Mater.* **2012**, *324*, 1891–1896. [CrossRef]
24. Armstrong, A.; Nilsén, F.; Rameš, M.; Colman, R.H.; Veřtát, P.; Kmječ, T.; Straka, L.; Müllner, P.; Heczko, O. Systematic Trends of Transformation Temperatures and Crystal Structure of Ni-Mn-Ga-Fe-Cu Alloys. *Shape Mem. Superelasticity* **2020**, *6*, 97–106. [CrossRef]
25. Kratochvílová, M.; Král, D.; Dušek, M.; Valenta, J.; Colman, R.; Heczko, O.; Veis, M. Fe_2MnSn—Experimental quest for predicted Heusler alloy. *J. Magn. Magn. Mater.* **2020**, *501*, 166426. [CrossRef]
26. Kratochvílová, M.; Klicpera, M.; Malý, F.; Valenta, J.; Veis, M.; Colman, R.; Heczko, O. Systematic experimental search for Fe_2YZ Heusler compounds predicted by ab-initio calculation. *Intermetallics* **2021**, *131*, 107073. [CrossRef]
27. Criss, J.W.; Birks, L.S. Calculation methods for fluorescent X-ray spectrometry. Empirical coefficients versus fundamental parameters. *Anal. Chem.* **1968**, *40*, 1080–1086. [CrossRef]

28. Rousseau, R. How to Apply the Fundamental Parameters Method to the Quantitative X-ray Fluorescence Analysis of Geological Materials. *J. Geosci. Geomat.* **2013**, *1*, 1–7. [CrossRef]
29. Haschke, M. Quantification. In *Laboratory Micro-X-ray Fluorescence Spectroscopy*; Springer International Publishing: Cham, Switzerland, 2014; Chapter 4, pp. 157–199. [CrossRef]
30. Kanomata, T.; Nunoki, S.; Endo, K.; Kataoka, M.; Nishihara, H.; Khovaylo, V.V.; Umetsu, R.Y.; Shishido, T.; Nagasako, M.; Kainuma, R.; et al. Phase diagram of the ferromagnetic shape memory alloys $Ni_2MnGa_{1-x}Co_x$. *Phys. Rev. B* **2012**, *85*, 134421. [CrossRef]
31. Straka, L.; Sozinov, A.; Drahokoupil, J.; Kopecký, V.; Hänninen, H.; Heczko, O. Effect of intermartensite transformation on twinning stress in Ni-Mn-Ga 10 M martensite. *J. Appl. Phys.* **2013**, *114*, 63504. [CrossRef]
32. Arrott, A. Criterion for Ferromagnetism from Observations of Magnetic Isotherms. *Phys. Rev.* **1957**, *108*, 1394–1396. [CrossRef]
33. Heczko, O.; Fähler, S.; Vasilchikova, T.M.; Voloshok, T.N.; Klimov, K.V.; Chumlyakov, Y.I.; Vasiliev, A.N. Thermodynamic, kinetic, and magnetic properties of a $Ni_{54}Fe_{19}Ga_{27}$ magnetic shape-memory single crystal. *Phys. Rev. B* **2008**, *77*, 174402. [CrossRef]
34. Graf, T.; Felser, C.; Parkin, S.S.P. Simple rules for the understanding of Heusler compounds. *Prog. Solid. State Chem.* **2011**, *39*, 1–50. [CrossRef]
35. Ramudu, M.; Kumar, A.S.; Seshubai, V.; Rajasekharan, T. Correlation of martensitic transformation temperatures of Ni- Mn-Ga/Al-X alloys to non-bonding electron concentration. *IOP Conf. Ser. Mater. Sci. Eng.* **2015**, *73*, 012074. [CrossRef]
36. Miedema, A.; de Châtel, P.; de Boer, F. Cohesion in alloys—Fundamentals of a semi-empirical model. *Physic B* **1980**, *100*, 1–28. [CrossRef]
37. Miedema, A.R.; de Boer, F.R.; de Chatel, P.F. Empirical description of the role of electronegativity in alloy formation. *J. Phys. F Met. Phys.* **1973**, *3*, 1558–1576. [CrossRef]
38. Nespoli, A.; Villa, E.; Passaretti, F.; Albertini, F.; Cabassi, R.; Pasquale, M.; Sasso, C.P.; Coïsson, M. Non-Conventional Techniques for the Study of Phase Transitions in NiTi-Based Alloys. *J. Mater. Eng. Perform.* **2014**, *23*, 2491–2497. [CrossRef]
39. Heczko, O.; Vronka, M.; Veřtát, P.; Rameš, M.; Onderková, K.; Kopecký, V.; Krátká, P.; Ge, Y. Mechanical Stabilization of Martensite in Cu–Ni–Al Single Crystal and Unconventional Way to Detect It. *Shape Mem. Superelasticity* **2018**, *4*, 77–84. [CrossRef]
40. O'Handley, R.C. *Modern Magnetic Materials: Principles and Applications*; John Wiley & Sons, Inc.: Hoboken, NJ, USA, 2000; ISBN 0-471-15566-7.
41. Straka, L.; Fekete, L.; Rameš, M.; Belas, E.; Heczko, O. Magnetic coercivity control by heat treatment in Heusler Ni–Mn–Ga(–B) single crystals. *Acta Mater.* **2019**, *169*, 109–121. [CrossRef]
42. Webster, P.J.; Ziebeck, K.R.A.; Town, S.L.; Peak, M.S. Magnetic order and phase transformation in Ni_2MnGa. *Phil. Mag. B* **1984**, *49*, 295–310. [CrossRef]
43. Vasil'ev, A.N.; Bozhko, A.D.; Khovailo, V.V.; Dikshtein, I.E.; Shavrov, V.G.; Buchelnikov, V.D.; Matsumoto, M.; Suzuki, S.; Takagi, T.; Tani, J. Structural and magnetic phase transitions in shape-memory alloys $Ni_{2+x}Mn_{1-x}Ga$. *Phys. Rev. B* **1999**, *59*, 1113–1120. [CrossRef]
44. Enkovaara, J.; Heczko, O.; Ayuela, A.; Nieminen, R.M. Coexistence of ferromagnetic and antiferromagnetic order in Mn-doped Ni_2MnGa. *Phys. Rev. B* **2003**, *67*, 212405. [CrossRef]
45. Kainuma, R.; Imano, Y.; Ito, W.; Sutou, Y.; Morito, H.; Okamoto, S.; Kitakami, O.; Oikawa, K.; Fujita, A.; Kanomata, T.; et al. Magnetic-field-induced shape recovery by reverse phase transformation. *Nature* **2006**, *439*, 957–960. [CrossRef]
46. Kainuma, R.; Oikawa, K.; Ito, W.; Sutou, Y.; Kanomata, T.; Ishida, K. Metamagnetic shape memory effect in NiMn-based Heusler-type alloys. *J. Mater. Chem.* **2008**, *18*, 1837–1842. [CrossRef]
47. Rameš, M.; Heczko, O.; Sozinov, A.; Ullakko, K.; Straka, L. Magnetic properties of Ni-Mn-Ga-Co-Cu tetragonal martensites exhibiting magnetic shape memory effect. *Scr. Mater.* **2018**, *142*, 61–65. [CrossRef]
48. Pushin, V.G.; Kourov, N.I.; Korolev, A.V.; Marchenkov, V.V.; Marchenkova, E.B.; Kazantsev, V.A.; Kuranova, N.N.; Popov, A.G. Effect of cobalt doping on thermoelastic martensitic transformations and physical properties of magnetic shape memory alloys $Ni_{50-x}Co_xMn_{29}Ga_{21}$. *Phys. Solid State* **2013**, *55*, 2413–2421. [CrossRef]
49. Söderberg, O.; Koho, K.; Sammi, T.; Liu, X.; Sozinov, A.; Lanska, N.; Lindroos, V. Effect of the selected alloying on Ni-Mn-Ga alloys. *Mater. Sci. Eng. A* **2004**, *378*, 389–393. [CrossRef]
50. Cherechukin, A.; Khovailo, V.; Koposov, R.; Krasnoperov, E.; Takagi, T.; Tani, J. Training of the Ni-Mn-Fe-Ga ferromagnetic shape-memory alloys due cycling in high magnetic field. *J. Magn. Magn. Mater.* **2003**, *258–259*, 523–525. [CrossRef]
51. Kumar, A.S.; Ramudu, M.; Seshubai, V. Effect of selective substitution of Co for Ni or Mn on the superstructure and microstructural properties of $Ni_{50}Mn_{29}Ga_{21}$. *J. Alloys Compd.* **2011**, *509*, 8215–8222. [CrossRef]
52. Soto, D.; Hernández, F.A.; Flores-Zúñiga, H.; Moya, X.; Mañosa, L.; Planes, A.; Aksoy, S.; Acet, M.; Krenke, T. Phase diagram of Fe-doped Ni-Mn-Ga ferromagnetic shape-memory alloys. *Phys. Rev. B* **2008**, *77*, 184103. [CrossRef]
53. Chen, X.Q.; Lu, X.; Wang, D.Y.; Qin, Z.X. The effect of Co-doping on martensitic transformation temperatures in Ni–Mn–Ga Heusler alloys. *Smart Mater. Struct.* **2008**, *17*, 065030. [CrossRef]
54. Belosludtseva, E.S.; Kuranova, N.N.; Marchenkova, E.B.; Popov, A.G.; Pushin, V.G. Effect of gallium alloying on the structure, the phase composition, and the thermoelastic martensitic transformations in ternary Ni–Mn–Ga alloys. *Tech. Phys.* **2016**, *61*, 547–553. [CrossRef]
55. Yang, S.; Wang, C.; Liu, X. Phase equilibria and composition dependence of martensitic transformation in Ni–Mn–Ga ternary system. *Intermetallics* **2012**, *25*, 101–108. [CrossRef]

56. Ingale, B.; Gopalan, R.; Rajasekhar, M.; Ram, S. Studies on ordering temperature and martensite stabilization in $Ni_{55}Mn_{20-x}Ga_{25+x}$ alloys. *J. Alloys Compd.* **2009**, *475*, 276–280. [CrossRef]
57. Pushin, V.; Kuranova, N.; Marchenkova, E.; Pushin, A. Design and Development of Ti–Ni, Ni–Mn–Ga and Cu–Al–Ni-based Alloys with High and Low Temperature Shape Memory Effects. *Materials* **2019**, *12*, 2616. [CrossRef] [PubMed]
58. Zhang, Y.; Li, Z.; He, X.; Huang, Y.; Xu, K.; Jing, C. Evolution of phase transformation and magnetic properties with Fe content in $Ni_{55-x}Fe_xMn_{20}Ga_{25}$ Heusler alloys. *J. Phys. D Appl. Phys.* **2018**, *51*, 075004. [CrossRef]
59. Khan, M.; Dubenko, I.; Stadler, S.; Ali, N. The structural and magnetic properties of $Ni_2Mn_{1-x}M_xGa$ (M=Co, Cu). *J. Appl. Phys.* **2005**, *97*, 10M304. [CrossRef]
60. Richard, M.L.; Feuchtwanger, J.; Allen, S.M.; O'handley, R.C.; Lázpita, P.; Barandiaran, J.M.; Gutierrez, J.; Ouladdiaf, B.; Mondelli, C.; Lograsso, T.; et al. Chemical order in off-stoichiometric Ni-Mn-Ga ferromagnetic shape-memory alloys studied with neutron diffraction. *Philos. Mag.* **2007**, *87*, 3437–3447. [CrossRef]
61. Porro-Azpiazu, J.M.; Barandiaran, J.M.; Chernenko, V.; Feuchtwanger, J.; Lazpita, P.; Perez-checa, A.; Velamazan, R.; Alberto, J. *Role of Fe and Co Addition in NiMnGa Shape Memory Alloys: Site Occupancy and Structural Stabilization of Crystallographic Phases*; Technical Report; Institut Laue-Langevin (ILL): Grenoble, France, 2018. [CrossRef]
62. Lazpita, P.; Barandiaran, J.M.; Chernenko, V.; Feuchtwanger, J.; Gutierrez, J.; Hansen, T.; Martinez, C.F.J.; Mondelli, C.; Perez-Checa, A.; Sozinov, A.; et al. *Influence Fe Addition on Structural and Magnetic Properties of $Ni_{45}Co_5Mn_{25-x}Fe_xGa_{20}Cu_5$ (x = 0, 1, 2, 4, 5, 6 and 8) Heusler Alloys*; Technical Report; Institut Laue-Langevin (ILL): Grenoble, France, 2018. [CrossRef]
63. Perez-Checa, A.; Lazpita, P.; Porro-Azpiazu, J.M.; Feuchtwanger, J.; Velamazan, J.A.R. *Site occupancy study of Fe doped $Ni_{45}Co_5Mn_{25-x}Fe_xGa_{20}Cu_5$ (x =4 and 5) Heusler single crystals*; Technical Report; Institut Laue-Langevin (ILL): Grenoble, France, 2018. [CrossRef]
64. Li, C.M.; Luo, H.B.; Hu, Q.M.; Yang, R.; Johansson, B.; Vitos, L. Site preference and elastic properties of Fe-, Co-, and Cu-doped Ni_2MnGa shape memory alloys from first principles. *Phys. Rev. B* **2011**, *84*, 024206. [CrossRef]
65. Jie, C.; Yan, L.; Jia-Xiang, S.; Hui-Bin, X. Site Preference and Alloying Effect of Excess Ni in Ni-Mn-Ga Shape Memory Alloys. *Chin. Phys. Lett.* **2009**, *26*, 047101. [CrossRef]
66. Ayila, S.K.; Machavarapu, R.; Vummethala, S. Site preference of magnetic atoms in Ni-Mn-Ga-M (M = Co, Fe) ferromagnetic shape memory alloys. *Phys. Status Solidi B* **2011**, *249*, 620–626. [CrossRef]

Article

Transitory Ultrasonic Absorption in "Domain Engineered" Structures of 10 M Ni-Mn-Ga Martensite

Sergey Kustov [1,*], Andrey Saren [2], Bruno D'Agosto [1], Konstantin Sapozhnikov [3], Vladimir Nikolaev [3] and Kari Ullakko [2]

[1] Departament de Física, Universitat de les Illes Balears, cra Valldemossa, km. 7.5, 07122 Palma de Mallorca, Spain; bruno.dagosto1@estudiant.uib.cat
[2] Material Physics Laboratory, School of Engineering Science, LUT University, Yliopistonkatu 34, 53850 Lappeenranta, Finland; andrey.saren@lut.fi (A.S.); kari.ullakko@lut.fi (K.U.)
[3] Shaped Crystals Physics Laboratory, Solid State Physics Division, Ioffe Institute, Polytechnicheskaya, 26, 194021 St. Petersburg, Russia; k.sapozhnikov@mail.ioffe.ru (K.S.); nkvlad@inbox.ru (V.N.)
* Correspondence: Sergey.Kustov@uib.es; Tel.: +34-971-171375

Abstract: In this work we create in 10 M Ni-Mn-Ga martensitic samples special martensitic variant structures consisting of only three twins separated by two *a/c* twin boundaries: Type I and Type II, with relatively low and very high mobility, respectively. The "domain engineered" structure thus created allows us to investigate the dynamics of a single highly mobile *a/c* twin boundary (TB). We show that temperature variations between 290 and 173 K in our samples induce an intense transitory internal friction at ultrasonic frequencies ca. 100 kHz, peaking around 215 K. A comparison is made of the data for the "domain engineered" sample with the behaviour of reference samples without *a/c* TB. Reference samples have two different orientations of *a/b* twins providing zero and maximum shear stresses in *a/b* twinning planes. We argue, first, that the transitory internal friction, registered at rather high ultrasonic frequencies, has magnetic origin. It is related with the rearrangement of magnetic domain structure due to the motion of *a/c* twin boundary induced by thermal stresses. This internal friction term can be coined "magnetic transitory internal friction". Magnetic transitory internal friction is a new category, linking the classes of transitory and magnetomechanical internal friction. Second, the structure of *a/b* twins is strongly non-equilibrium over a broad temperature range. As a consequence, the Young's modulus values of the samples with maximum shear stress in *a/b* twinning planes can take any value between ca. 15 and 35 GPa, depending on the prehistory of the sample.

Keywords: magnetic shape memory alloys; twinning; *a/c* twins; *a/b* twin laminate; domain engineering; internal friction; Young's modulus

1. Introduction

10 M Ni-Mn-Ga martensites are examples of a multiferroic material in which strong coupling between ferroic subsystems—elastic and magnetic—is responsible for such useful property as magnetic actuation. Here we observe and describe a new phenomenon, associated with local magnetic flux change provoked by the motion of an elastic domain wall—*a/c* twin boundary (TB). This new phenomenon is the ultrasonic transitory internal friction (IF) of magnetic origin which is a new category of damping, linking the classes of transitory and magnetomechanical IF. In our experiments we use the "domain engineered" 10 M Ni-Mn-Ga samples with controlled structure of *a/c* and *a/b* twins. Such experiments reveal new striking features of apparent elastic properties of *a/b* twin laminate.

1.1. Transitory Internal Friction

The notion "Transitory internal friction" refers to the component of absorption of mechanical energy of periodic vibrations that emerges when oscillatory stress is coupled with a certain external field. The superimposed coupling field results in an extra irreversible strain produced by periodic stress and, hence, additional damping of oscillations. The irreversibility of strain marks the transitory IF, IF_{tr}, since in conventional structural internal friction experiments anelastic strain is usually reversible. Physical origins of the additional periodic irreversible strain and, hence, of coupling fields can be different:

- first order structural phase transition in a solid [1]; this transition can be induced by temperature, mechanical stress or any other relevant parameter such as magnetic field;
- plastic or even microplastic deformation of crystals during active deformation, creep or temperature variation in non-cubic polycrystals [2]; external quasistatic stress is coupled then with the oscillatory stress.

The general formal description of the IF_{tr} remains essentially the same despite apparent difference in potential specific physical mechanisms. The transitory internal friction is a strong function of strain-coupled external field and parameters of the measuring procedure, such as heating–cooling (stress, magnetic field, etc.) rate, frequency of measurements and oscillatory strain amplitude. During plastic or microplastic deformation, in a first approximation, IF_{tr} is expressed as [2]:

$$IF_{tr} = K\frac{\dot{\varepsilon}}{f\varepsilon_0}, \tag{1}$$

where K is a constant, $\dot{\varepsilon}$ is the strain rate, f and ε_0 are the frequency and elastic strain amplitude of mechanical oscillations, respectively.

The first order structural transition is accompanied by lattice strain. In the case of first order structural transition induced, for example, by temperature variation, the strain rate in Equation (1) is substituted for temperature variation rate or T-dot, $dT/dt \equiv \dot{T}$:

$$IF_{tr} = K'\frac{\dot{T}}{f\varepsilon_0}, \tag{2}$$

with K'—numerical factor.

Equations (1) and (2) are easily interpreted qualitatively: IF_{tr} is proportional to transformation or deformation rate, since the amount of additional strain per period of oscillations is proportional to these rates. The same reason explains the inverse frequency dependence: the higher the frequency, the lower the additional strain accumulated during one period of oscillations under constant strain rate or T-dot. The inverse dependence on strain amplitude stems from the definition of the internal friction as proportional to the ratio between the energy dissipated in a cycle of oscillations ΔW and maximum elastic stored energy W:

$$IF \propto \frac{\Delta W}{W}. \tag{3}$$

Since the denominator in the ratio, Equation (3), $W \propto \varepsilon_0^2$, a simple assumption that ΔW is proportional to the elastic strain amplitude yields the inverse proportionality of IF_{tr} to ε_0. Many specific refinements of this simple interpretation have been suggested, see, e.g., reviews [1,2]. An important property of IF_{tr} is the inverse (or nearly inverse in more refined theories [1]) frequency dependence, Equations (1) and (2). Therefore, it is generally accepted that the IF_{tr} can be observed only at low frequencies, $f < 10^3$ Hz, and that it takes negligible values at higher, ultrasonic frequencies [1–3].

One of the conditions under which the IF_{tr} emerges is a variation in temperature of polycrystalline materials with non-cubic lattice that possess anisotropy of thermal expansion coefficients, see, e.g., Ref. [2] and references therein. The anisotropy of thermal expansion is responsible for thermal stresses generated in differently oriented grains when the temperature of the polycrystalline aggregate changes. These thermal stresses induce microplastic or even plastic deformation of grains and, hence, the transitory IF emerges [2]. Under these circumstances, the IF_{tr} is proportional to the cooling–heating rate, T-dot. For the same reason, thermal stresses are induced in multiphase materials, for example, precipitation hardened [3] and composite materials [4]. Another group of materials showing transitory IF during thermal cycling are shape memory alloys in the martensitic phase. The martensite is usually formed by self-accommodating twin-related elastic domains ("martensitic variants") with low lattice symmetry and high anisotropy of thermal expansion [5]. Therefore, similar to polycrystalline aggregates, martensitic phases show the transitory IF, even if the material is single crystalline in the high-temperature phase. IF_{tr} was observed and studied in the martensitic phase of classic Cu-Al-Ni [6] and NiTi [7] shape memory alloys. IF_{tr} in martensites provides information on generation of thermal stresses, dynamics of twin boundaries, their pinning, etc.

1.2. Magnetomechanical Internal Friction

The magnetomechanical IF or magnetomechanical damping (MMD) is defined as any damping suppressed by a magnetic field [8]. The classic description of MMD considers three additive components, associated with magnetic domain walls and net magnetization of the sample. Two MMD components are linear in strain (with the IF value independent of strain amplitude ε_0) and one—non-linear—whose value is strain amplitude-dependent [8–10]. Expressed as logarithmic decrement of oscillations, δ, these components are: a linear microeddy current damping, δ_μ, usually found for low strain amplitudes; a non-linear hysteretic damping, $\delta_h(\varepsilon_0)$, emerging at higher strain amplitudes, and a macroeddy current damping component, δ_M, which is also linear in strain amplitude or strain amplitude independent. δ_μ and $\delta_h(\varepsilon_0)$ are ascribed to a short-range reversible and long-range hysteretic motion of magnetic domain walls, respectively. Macroeddy MMD originates from macroscopic net magnetization of the sample. Although by definition all three MMD terms vanish at saturation, they differ in their dependence on field. δ_μ and $\delta_h(\varepsilon_0)$ attain maximum values for the demagnetized state of a sample (magnetization $M = 0$) then decline monotonously with applied field H, see e.g., Refs. [8–11], whereas $\delta_M = 0$ when $M = 0$. Therefore, $\delta_M (H)$ has a maximum for H somewhat above half saturating field. Thus, only microeddy damping is relevant at low oscillation amplitudes in the demagnetized samples, without external field. A crucial property of microeddy MMD is the linear frequency dependence up to a frequency somewhat lower than the frequency of magnetic domain wall relaxation with typical values between 100 kHz and 10 MHz [10,12]. This frequency dependence is a consequence of Faraday's law controlling the intensity of the energy dissipation by eddy currents.

Recently, T-dot dependent IF was reported in polycrystalline Dy below the Néel temperature, in the ordered antiferromagnetic state [13] at an ultrasonic frequency close to 90 kHz. Since the T-dot effect was observed only in the magnetically ordered state, it had been attributed to the magnetic domain structure; more specifically, to the magnetic domain walls perpendicular to the c-axis (axis of the sixth order in the hexagonal structure), which carry magnetic moments. The possibility to observe T-dot effect at an ultrasonic frequency was associated in [13] with the presumably magnetic origin of losses, which are proportional to the frequency in the low-amplitude range. This proportionality compensates the inverse frequency dependence of the transitory IF, Equation (2), and makes the IF_{tr} essentially frequency-independent up to a frequency approaching the frequency of domain wall relaxation. Taking the relaxation frequency ca. 100 kHz–10 MHz, one gets the range of nearly frequency-independent transitory MMD up to 10 kHz–1 MHz [14].

In this work we intended to test whether the IF during thermal cycling of ferromagnetic martensites shows the magnetic ultrasonic transient internal friction. The idea behind this test is quite simple:

- martensitic alloys consisting of several differently oriented elastic domains (martensitic variants) show transitory IF due to the rearrangement of twin structure [6,7];
- in ferromagnetic martensites magnetic domain structure and structure of martensitic variants are strongly coupled [15–17] due to a high magnetocrystalline anisotropy; hence, variations of martensitic variant structure due to thermal stresses will inevitably

produce rearrangement of the magnetic domain structure and, probably, magnetic transitory IF term.

It has to be emphasized that the transitory IF related to a change in magnetic domain structure can be classified as a new category of both transitory and magnetomechanical IF.

2. Materials and Methods

2.1. Material

For several reasons Ni-Mn-Ga ferromagnetic martensitic alloy with layered structure, (so-called 5 layered or 10 M martensite) with slightly monoclinically distorted tetragonal lattice was chosen. First, this type of the martensite is known to possess rather low twinning stress: 0.5–1.5 MPa for polyvariant samples and below 0.1 MPa for a single Type II twin boundary [18–20]. Therefore, it was expected to suffer, during temperature variations, microplastic strains sufficient to produce transitory effects. Second, the structure of this martensite is well documented due to its potential in practical applications. Recent studies show that 10 M martensite possesses an hierarchical 4-level twin structure [21,22]. a/c twins, observable at a macroscopic level, can have dimensions comparable with the sample size. a/c twins are responsible for the so-called magnetostrain: large strains that can be induced in 10 M martensite by an external magnetic field [23,24]. Internal twins inside a/c twins are mesoscopic level modulation twins with (100) and (010) type twinning planes. The small difference between a and b lattice parameters of the order of 0.002 nm [25] is nevertheless sufficient to create, at a microscopic level, a/b twins with (110) type twinning planes. Their thickness ranges from hundreds of nanometres to tens of microns [26–28]. The finest structural level is associated with so-called adaptive nanotwins of 2, 3 or 5 atomic layers thick. The difference between a, b and c axes results in a substantial anisotropy of thermal expansion of the variants joint by a/c twin boundaries [5]. Since c-axis is the direction of easy magnetization, any displacement of an a/c twin boundary (TB) provokes local variations of the magnetic flux. Two types of a/c TBs were found in 10 M Ni-Mn-Ga martensite: Type I with (1 0 1) twinning planes and Type II with irrational indices close to (10 1 10). Type I TBs have much lower mobility (twinning stress around 0.5–1 MPa) as compared to Type II ones (twinning stress below 0.1 MPa) [20].

2.2. Experimental Method and Protocol

The piezoelectric ultrasonic composite oscillator technique (PUCOT) [29,30], was used to measure elastic and anelastic (internal friction) properties of the samples at a frequency close to 90 kHz. The PUCOT employs longitudinal resonant oscillations of the oscillator, consisting of two quartz transducers and a sample attached to them. A fully automated experimental setup (see ref. [30] for a more detailed description) allowed us to measure and control the strain amplitude of the oscillations, register the values of the internal friction, expressed as logarithmic decrement, δ, and resonant frequency of the longitudinal oscillations of the sample, f. Strong variations of the Young's modulus (YM), and, hence, of the resonant frequency of the samples, were found for certain configurations of a/b twins. Therefore, instead of the approximate solution for the resonant frequency of samples [30], a precise one [31] was used:

$$m_q f_q \tan \frac{\pi f_c}{f_q} + m_s f \tan \frac{\pi f_c}{f} = 0 \qquad (4)$$

where m_q and m_s are the masses of the quartz transducer and of the sample, f_q and f_c are the resonant frequencies of the quartz transducer alone and with the sample attached, respectively.

The YM of the sample E is determined from the resonant frequency f, length of the sample l and its density ρ:

$$E = 4\rho \frac{f^2 l^2}{n^2}, \qquad (5)$$

where $n = 1, 2, 3 \ldots$ is the harmonic number.

Three experimental protocols employed consisted in (i) measurements of δ and f as a function of strain amplitude in order to determine the ranges of amplitude-independent and amplitude-dependent damping; these ranges correspond to predominantly linear and non-linear dynamics of twins under periodic stress in acoustic experiments; (ii) temperature cycling the sample between 290 and 173 K under a constant T-dot; (iii) interruptions of cooling/heating scans in several points of the thermal cycles and registration of the kinetics of relaxation of the IF and frequency under isothermal conditions during 2400 s. Measurements of temperature spectra were performed at low strain amplitudes, $\varepsilon_0 = 2 \times 10^{-7}$, in order to avoid non-linearity of the IF provoked by a large scale motion of elastic and magnetic domain walls. Cooling/heating rate during temperature spectra measurements was 1 or 2 K/min.

2.3. Samples

The samples produced by AdaptaMat Ltd., Finland, had the composition $Ni_{50.0}Mn_{28.4}Ga_{21.6}$ (± 0.3 at. %) and martensitic transformation temperatures slightly above 300 K [21]. The samples were bar-shaped, approximately $1.0 \times 1.0 \times (11\text{–}12)$ mm^3, with surfaces parallel to {100} planes of the high-temperature cubic phase. The length of the sample was chosen to fit best the resonant frequency of the quartz transducer at room temperature. The surface of the samples was mechanically and chemically polished. Three different configurations of a/c and a/b twins were created in the samples, which are shown schematically in Figure 1. The major part of the experiments was performed for the so-called "domain engineered" [32] Structure 1 sample that contained three martensitic variants with differently oriented c-axes, separated by two a/c TBs of different types—see Figure 1a.

Two additional configurations of twins were checked (Structure 2, Figure 1b, and Structure 3, Figure 1c), similar to those studied in Ref. [33]. Both of these configurations did not contain a/c TBs and had easy magnetization c-axis parallel (Structure 2, Figure 1b) or perpendicular (Structure 3, Figure 1c) to the sample long side. Therefore, the shear stresses in the a/b twinning planes were absent or maximum, respectively, for these two configurations of a/b twins. Details of the preparation of samples with Structure 2 and Structure 3 of twins are reported in [33].

An optical image in Figure 2 taken in polarized light shows the central part of the sample with Structure 1, clarifying the a/c twin configuration schematically depicted in Figure 1a. The narrow variant with the c-axis perpendicular to the sample long side is always situated in the middle part of the sample. Two big variants with c-axis parallel to the sample length occupy both sides of the sample. The three variants are separated by two a/c twin boundaries: one with low mobility, Type I, and another one of high mobility, Type II. The central section of the sample is close to the stress antinode and the shear stress in the plane inclined 45° with respect to the sample axis reaches maximum values in the narrow central variant. It is important to mention that a/b twinning planes are parallel to the sample long axis in the two big variants at the ends of the sample and do not experience shear stress, whereas a/b twins in the central narrow variant are under maximum shear stress and might contribute, apart from the a/c twins, to the measured elastic and anelastic effects. Traces of micron-scale a/b twins are not seen in Figure 2, since their observation in an optical microscope requires higher magnifications and is difficult in general because of their low optical contrast [28].

The additional twin configurations with Structure 2 and Structure 3, Figure 1b,c, served to clarify the role of a/c and a/b twins in the elastic and anelastic effects found during thermal cycling of the samples. For the orientation of samples used, modulation twins with twinning planes (100) and (010) type, if they exist, do not experience shear stress. Therefore, their contribution to elastic and anelastic effects can be discarded.

The effect of a/c and a/b twin configurations on the IF damping and YM spectra was checked for three different samples which showed similar results.

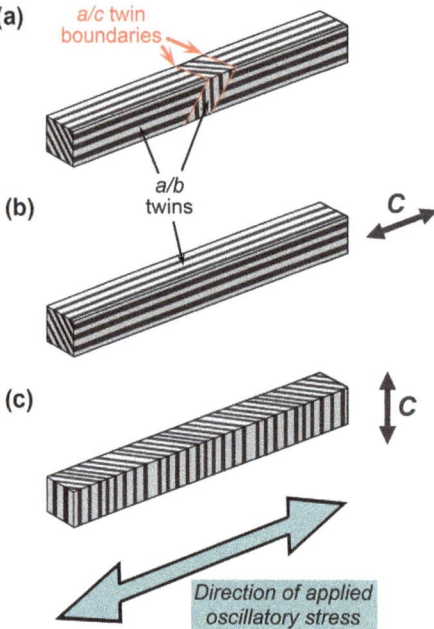

Figure 1. Schematic images of the *a/c* and *a/b* twin configurations in three types of the samples studied: (**a**) Structure 1 with two *a/c* twin boundaries, Type I and Type II, separating three martensitic variants; *a/b* twinning planes are parallel to the sample long axis in the two lateral variants and inclined 45° in the narrow central variant; (**b**) Structure 2 without *a/c* twin boundaries; *c*-axis and *a/b* twinning planes are parallel to the sample axis; (**c**) Structure 3 without *a/c* twin boundaries; *c*-axis is perpendicular to the sample axis and *a/b* twinning planes are inclined 45° with respect to the sample axis.

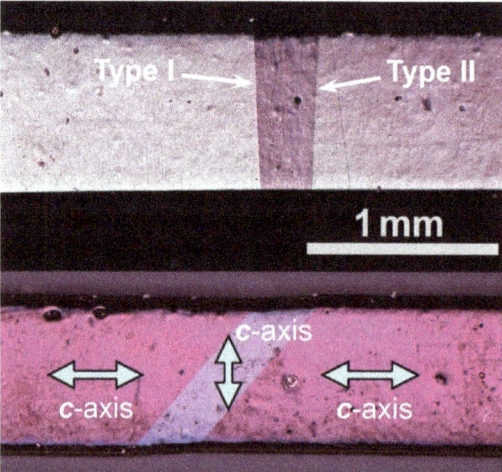

Figure 2. Two polarized light optical images of adjacent lateral surfaces of one of the studied samples with Structure 1 (Figure 1a). The sample consists of three martensitic variants separated by two *a/c* twin boundaries, Type I and Type II. The orientations of the *c*-axes, perpendicular in the two adjacent variants, are indicated by arrows.

3. Results

3.1. Strain Amplitude Dependence

Figure 3 shows the effect of oscillatory strain amplitude on the IF at room temperature. The strain amplitude dependences were registered in two consecutive scans as increasing and decreasing the oscillatory strain amplitude. The IF shows a non-linear pattern typical for different ferroelastics, including shape memory alloys [34–37]:

- a very weak non-linearity at low strain amplitudes, below ca. 6×10^{-7} for the increasing strain amplitudes, which transforms into a very steep increase in the IF;
- the high-amplitude part of the dependence is jerky; it shows abrupt changes of the IF and of the strain amplitude;
- a strong difference between direct run (increasing strain amplitudes) and reverse run (decreasing strain amplitude); we will refer to this effect as strain amplitude hysteresis;
- the strain amplitude hysteresis is well reproducible in two consecutive strain amplitude scans: it fully recovers in the second scan.

All these features are well established marks of the IF due to the TBs pinned by agglomerations of partially mobile pinning points (Cottrell-like clouds) [34–37]. The low amplitude stage of the strain amplitude dependence corresponds to the motion of TBs within pinning clouds. The sharp increase in the IF at the high-amplitude stage reflects the depinning of TBs from the clouds and the transition to the motion of TBs in the homogeneous spatial distribution of obstacles outside the Cottrell-like clouds. The amplitude hysteresis implies, within this model, the redistribution of pinning clouds by TBs moving in acoustic experiments. The reproducibility of the strain amplitude hysteresis indicates that the relaxation time of perturbed Cottrell-like clouds is less than the time to register a strain amplitude dependence (ca. 100 s).

Based on the strain amplitude scans, the low strain amplitude of 2×10^{-7} was selected for temperature spectra measurements in order to avoid the depinning of TBs by oscillatory stresses.

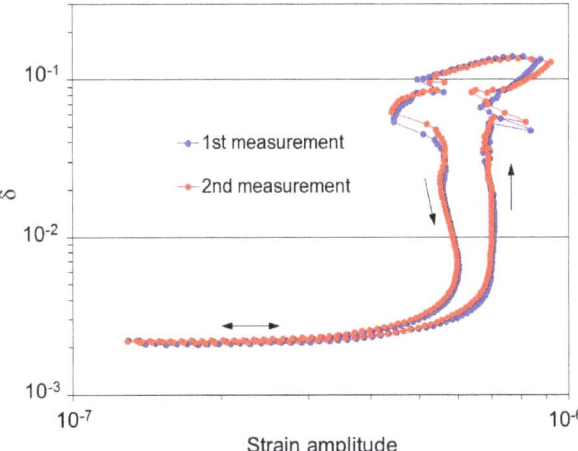

Figure 3. Strain amplitude dependence of the internal friction for a Structure 1 sample with one highly mobile (Type II) and one low mobility (Type I) a/c twin boundaries. Two consecutive measurements at room temperature (290 K), arrows indicate the scans with increasing and decreasing strain amplitude.

Figure 3 yields the critical strain amplitude for depinning of TBs and their jerky motion as $\varepsilon_0^{cr} \approx 8 \times 10^{-7}$. Taking the Young's modulus E of the sample Structure 1 as ca. 40 GPa (the Young's modulus of Structures 1–3 will be introduced later on), one obtains an estimation of the critical depinning stress, corresponding to the start of the jerky motion of TB as $\sigma_0^{cr} = E\varepsilon_0^{cr} \approx 0.03$ MPa. This value agrees well with the critical stress for the motion of a single Type II TBs [19,20], whereas movement of Type I TBs require more than order of magnitude higher stresses [20]. The non-linear IF in the same material due to *a/b* TB motion is a smooth power-law function of strain amplitude and does not show strain amplitude hysteresis [33]. Hence, the intense strain amplitude hysteresis and jerky non-linear anelastic behaviour point to the predominant role, at least in non-linear anelastic effects, of the highly mobile Type II *a/c* TB. An important conclusion relevant to dynamics of *a/c* TBs in 10 M Ni-Mn-Ga martensite is, thus, their interaction and pinning by Cottrell-like clouds of obstacles, which has not been detected for *a/b* TBs.

3.2. Temperature Dependence

Figure 4 shows temperature spectra of the IF registered for a Structure 1 sample during cooling and heating between 175 and 290 K. The crucial feature of the spectra is a broad maximum around 215 K. The peak shows a small temperature shift and also a minor difference in peak height between cooling and heating scans. Another effect is a strong abrupt increase in the IF during initiation of cooling from 290 K and a sharp drop at the lowest temperature of the thermocycle during the change from cooling to heating. These sharp variations in the IF level are indicative of the transitory, T-dot effect. To confirm this conclusion, several thermal cycles were performed with programmed interruptions of cooling/heating scans at several temperatures and isothermal exposures of the sample during 40 min during each interruption. An example of time dependence of temperature in one of the scans with interrupted cooling/heating is shown in Figure 5.

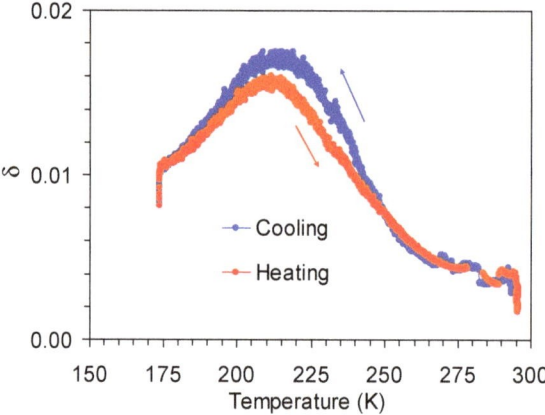

Figure 4. Continuous temperature spectra of the internal friction registered in a cooling–heating scan between 290 and 173 K for a Structure 1 sample with two *a/c* twin boundaries. Cooling–heating rate 1 K/min, oscillatory strain amplitude 2×10^{-7}.

Figure 6 exemplifies the results of one of the cooling scans interrupted at several temperatures for 2400 s. Experiments show an intense drop in the IF during each interruption of cooling/heating. The drop is especially intense, nearly tenfold, in the vicinity of the IF maximum. The green line connecting the values of the IF at the end of each isothermal holding represents the isothermal temperature spectrum which does not show any peak. It is also worth mentioning that the temperature spectra become noisy and the IF shows instabilities close the temperature of the maximum.

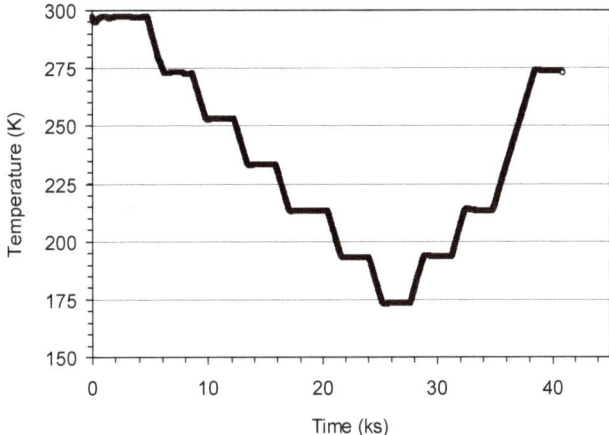

Figure 5. Example of the time dependence of temperature in one of the cooling–heating scans that was interrupted for 2400 s at 273, 253, 233, 213 and 193 K on cooling and at 193, 213 and 273 K on heating.

Kinetics of the IF during interruptions of cooling/heating allows one to identify the origin of the strong IF decline upon interruption of cooling. Several possible scenarios can be foreseen:

- the decline is time dependent;
- the decline is T-dot dependent;
- the decline is a combination of the two abovementioned processes, both time and T-dot dependent.

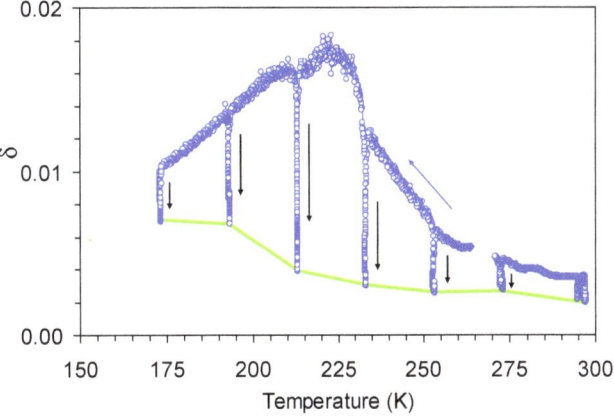

Figure 6. Temperature dependence of the internal friction for a Structure 1 sample with two a/c twin boundaries in a cooling scan between 290 and 173 K interrupted for 2400 s at 273, 253, 233, 213, 193 and 173 K, as shown in Figure 4. Cooling rate 1 K/min, strain amplitude 2×10^{-7}. Green line demonstrates the temperature spectrum under isothermal conditions. Vertical arrows mark isothermal decrease in the internal friction with time during interruptions of cooling.

These scenarios can and will be distinguished based on the IF kinetics during interruptions of cooling/heating. However, even instabilities of temperature control provide a spectacular evidence of the intense T-dot dependence of the IF during thermal cycling of the sample with a/c TBs. Figure 7 shows the IF in a cooling/heating cycle between 290 and 163 K at 2 K/min. The temperature control was not established down to 220 K during initial filling of the cryostat with liquid nitrogen. The temperature scan interruptions were programmed for 226 K on cooling and for 200 K on heating. Figure 7a indicates the instabilities of the IF during the initial period of the scan followed by a conventional IF peak centred around 215 K. Figure 7b shows the temperature, the absolute value of its time derivative dT/dt and the IF during heating scan. A gradual IF decrease with time is clearly seen during isothermal holding at 200 K. This overall trend is combined with a fast IF decline at the moment of the scan interruption. The most indicative is the existence of a sharp minimum that coincides with the zero of dT/dt which is provoked by overshoot during temperature stabilization. This fact proves the existence of the transitory IF term, proportional to T-dot. Figure 7c shows similar time dependences registered during interruption of cooling at 226 K under poor temperature control. The overall temperature variation during this interruption was ca. 4 K, between 224 and 228 K. No correlation between temperature variation during interruption of cooling and the IF behaviour is seen. On the other hand, excellent correlation is found between the IF and the absolute value of T-dot, Figure 7c. Thus, the time dependence under isothermal conditions, clearly seen in Figure 7b, and T-dot dependence point to the coexistence of time- and T-dot dependent IF terms in the temperature spectra of the sample with three martensitic variants.

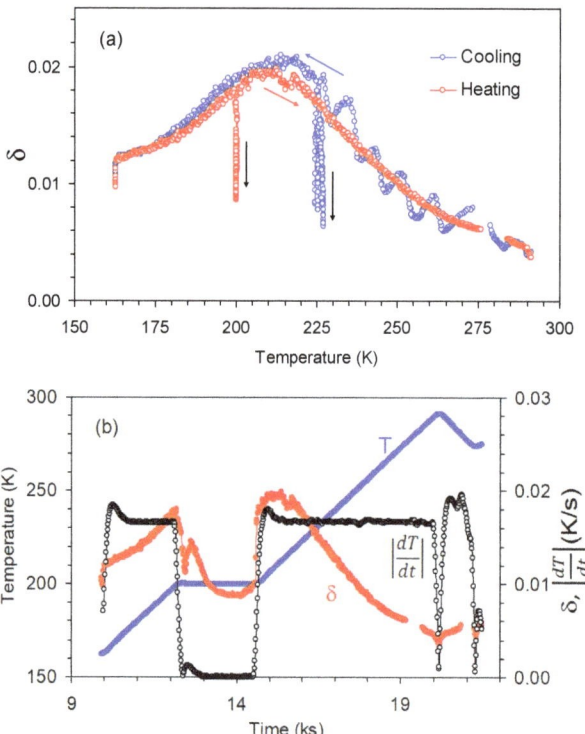

Figure 7. *Cont.*

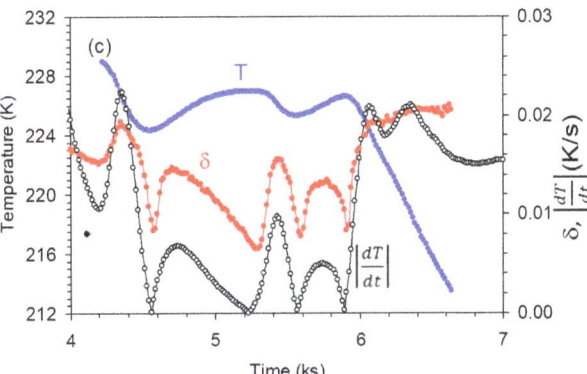

Figure 7. Temperature dependence of the internal friction for a Structure 1 sample with two a/c twin boundaries in a thermal cycle between 290 and 163 K, interrupted for 2400 s during cooling at 226 K and during heating at 200 K (**a**) and corresponding time dependences of temperature T, absolute value of cooling/heating rate dT/dt and the internal friction δ during the entire heating scan (**b**) and during interruption of cooling (**c**).

3.3. T-Dot Dependence

Dependences of the IF versus T-dot during interruptions of temperature scans provide more detailed information on the relative role of time- and T-dot-dependent terms in the IF spectrum [13]. Such dependences are depicted in Figure 8 for a Structure 1 sample with a/c TBs for interruptions of cooling at 233 and 193 K. The experimental results follow the same trend as has been reported in polycrystalline Dy [13] and interpreted as due to a superposition of several IF terms. It is convenient to separately analyse the dependences of the IF during the decrease in T-dot and isothermal holding and during re-initiation of cooling. During decrease in T-dot upon interruption of cooling the IF shows a linear decline, proportional to T-dot, reflecting the transitory IF term. The linear slope registered at 233 K is higher than at 193 K, in agreement with the temperature dependence of the transitory IF, Figure 6. The deviation from this linear trend appears at low T-dot values and transforms into the time-dependent isothermal drop of the IF under $dT/dt = 0$. After re-initiation of cooling, the IF first remains nearly constant and recovers the proportionality to T-dot after certain overcooling (by ca. 0.3 K). Thus, the IF kinetics during interruptions of cooling/heating proves the existence of a "fast", T-dot dependent, and "slow", time-dependent components. Figures 7b and 8 reproduce in detail the pattern reported for polycrystalline Dy [13]. Therefore, we will follow the same phenomenological qualitative interpretation elaborated for antiferromagnetic domain walls in Dy. Assuming additivity (linear independence) of different IF terms, the overall registered IF can be written in the following form [13]:

$$\delta\left(T, \left|\dot{T}\right|, t\right) = \delta_0(T) + \delta_{slow}(T, t) + \delta_{fast}\left(T, \left|\dot{T}\right|\right), \tag{6}$$

where $\delta_0(T)$ is the isothermal temperature spectra, similar to the one shown by green line in Figure 6, $\delta_{slow}(T, t)$ and $\delta_{slow}\left(T, \left|\dot{T}\right|\right)$ are the "slow" time-dependent and "fast" T-dot dependent terms. Figure 8 shows the separation of the "fast" and "slow" IF components in the total IF. These results indicate, in agreement with the data in Figure 6, that the isothermal structural term $\delta_0(T)$ represents only minor part of the overall IF.

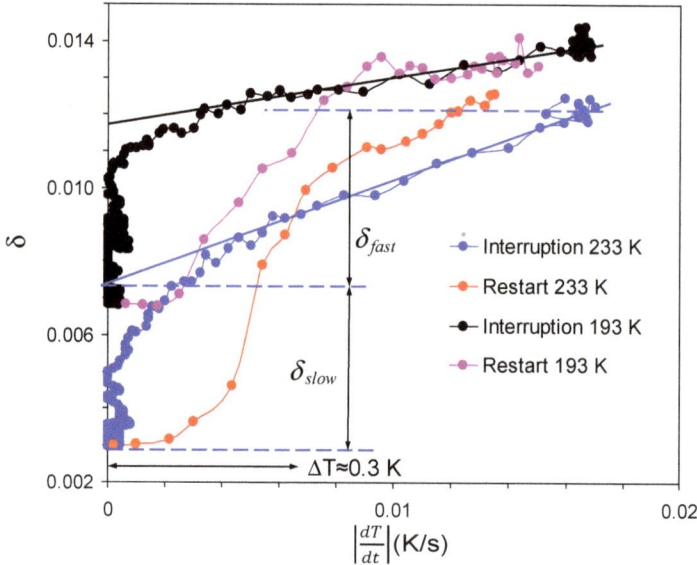

Figure 8. Internal friction versus absolute value of T-dot during interruption of cooling at 233 and 193 K of Structure 1 sample with two a/c twin boundaries. Vertical arrows mark the "fast" and "slow" internal friction components δ_{fast} and δ_{slow} in the data taken at 233 K.

4. Discussion

4.1. Comparison with Previous Results

To begin with, we compare our observations of the IF peak around 215 K with existing reports on Ni-Mn-Ga martensites we are aware of [38–42]. Previous experiments were mostly performed at low frequencies using Dynamic Mechanical Analysers (DMA). Unfortunately, the data available are scarce and certain parameters, such as oscillatory strain amplitude are not or poorly defined. Normally, the strain amplitudes employed in DMA tests are close to 10^{-4}. This strain amplitude falls within the range of strongly non-linear IF in martensitic Ni-Mn-Ga [33,43]. Due to the unknown contribution of non-linear effects in the low-frequency DMA data, the direct comparison of the absolute values of the IF registered at low frequencies and in our experiments is not feasible. Still, one concludes that the overall IF levels, and especially the transitory terms, are comparable at low and ultrasonic frequencies. Even though the strain amplitude used in the present investigation was ca. 10^3 times lower than in the DMA, the frequency employed was 10^5 times higher. Equation (2) shows that under these conditions, the conventional transitory IF term should be totally negligible in our ultrasonic experiments.

Certain previous observations and conclusions can be confronted qualitatively with the present results. We mention first that the IF peak observed is not related to intermartensitic transitions, since it does not show expected wide temperature hysteresis and, therefore, is different from the maxima reported by Chernenko et al. [38]. Nevertheless, the maximum found in the martensite of $Ni_{51.2}Mn_{31.1}Ga_{17.7}$ around 280–300 K [38] might have the same origin as the one found in the present study. Liu et al. [39] observed an IF peak close to 220 K in a $Ni_{51.5}Mn_{23.5}Ga_{25}$ alloy with the same sequence of phase transitions as in our single crystal. They classified this peak as a "twin boundary motion peak", which showed a strong decrease with a frequency between 0.1 and 1.0 Hz. The IF peak was hardly detectable at 1 Hz and, if existent, much less sensitive to the frequency between 1 and 5 Hz. Chang and Wu [40] reported an IF maximum around 215 K in a $Ni_{52.0}Mn_{23.5}Ga_{21.5}$ alloy belonging to the same family as present single crystals. Similar to the present observations, the peak apparently disappeared in the isothermal spectrum, although this property has not been commented upon. The authors described this peak as relaxation associated with twin boundaries. However, this interpretation contradicts frequency independence of the peak position and its disappearance in the isothermal spectrum. The kinetics of the IF relaxation were studied by Chang and Wu only over the ranges of phase transitions. They did not register temperature variations in parallel with damping and attributed the time dependence exclusively to the transitory IF term, despite very long-term kinetics, especially in $Ni_{53.8}Mn_{26.8}Ga_{19.4}$ from the Group III of Ni-Mn-Ga alloys. The results reported in [38–40] are, thus, compatible with our observations.

Data by Wang et al. [41] did not show an IF peak over this temperature range in a similar $Ni_{52}Mn_{24}Ga_{24}$ alloy in the demagnetized state. However, the peak emerged under applied field of 0.4 T and showed very weak decline with frequency. It is worth mentioning that Wang et al. [42] reported an additional nearly frequency-independent broad peak in the 14M martensite centred around 230 K for a quaternary $Ni_{52}Mn_{16}Fe_8Ga_{24}$ alloy. The authors could not suggest any interpretation of the maximum uncovered. We note that a crucial property of the peak was its temperature shift with strain amplitude. The maximum also marked a notable change in the effect of strain amplitude on damping temperature spectra. This property agrees with the important role of pinning process in the twin boundary mobility revealed in the present experiments. The change of the type of the strain amplitude dependence upon crossing the peak temperature in low-frequency experiments of Ref. [42] is easily explained by a change from dragging (high temperatures) to depinning (low temperatures) of twin boundaries during immobilization of obstacles in a cooling scan [35,37]. To summarize this brief review, we can conclude that our results agree with a number of previous reports [38–40]. Still, observations and interpretations of the IF and elastic properties of Ni-Mn-Ga modulated structures remain controversial. The use of samples with unknown/uncontrolled twin configuration is a serious drawback of previous works. The novelty of the present investigation consists in detailed study of both time and T-dot dependences of the IF and, most importantly, of well controlled "engineered" structures of both *a/c* and *a/b* twins.

4.2. Role of a/c and a/b Twins in the Internal Friction and Young's Modulus Temperature Spectra

The data for samples with three different types of *a/b* and *a/c* twin structure clarify the contribution of these two types of twins to the IF and YM temperature spectra of 10 M Ni-Mn-Ga. The sample with twin structure shown in Figure 1b does not have *a/c* TBs and the shear stress acting in *a/b* twinning planes essentially does not exist (c-axis is aligned along the sample). The sample of Figure 1c does not have *a/c* TBs either, but the shear stress in the *a/b* twinning planes is maximum (c-axis is perpendicular to the long side of the sample). Therefore, a comparison of the results for the structures in Figure 1a,b is expected to yield the contribution of the *a/c* TB and of *a/b* twins in the narrow central part of the "domain engineered" sample. The sample with maximum *a/b* shear stress should clarify the potential contribution of *a/b* TBs of the thin central variant of the sample in Figure 1a.

The temperature spectra of the YM and IF for the samples with three types of twin structure are shown in Figure 9. The Structure 2 sample (Figure 1b) shows the highest value of the YM with hardly detectable thermal hysteresis, Figure 9a, and extremely low IF values, comparable with background damping of the quartz transducers, Figure 9b. No transitory IF peak is found in absence of *a/c* TBs.

The effect of temperature on the IF and YM of the Structure 3 sample is totally different and perfectly agrees with the results of Ref. [33]. The YM behaviour is highly anomalous, Figure 9a: (i) it shows an unusually high hysteresis between cooling and heating scans; (ii) the YM temperature spectra show several intermittent stages of normal and anomalous temperature dependence; (iii) the Young's modulus behaviour is strongly history-dependent, especially for temperatures exceeding 260–270 K, where the modulus drops rapidly with temperature. Interestingly enough, the modulus declines strongly on heating above 275 K, but this decline is not reversed and continues on subsequent cooling, Figure 9a. The YM can take, over a wide temperature range, the values between ca. 35

and 10 GPa, depending on the previous history of the sample. As an indicative number, the change of the highest temperature of the cycle from 284 and 292 K results in a twofold difference of the YM values on further cooling, Figure 9a. It is remarkable that, despite high variability of the YM values, the complicated pattern of the IF and YM temperature spectra is perfectly reproducible provided the highest and the lowest temperatures of the cycles are the same.

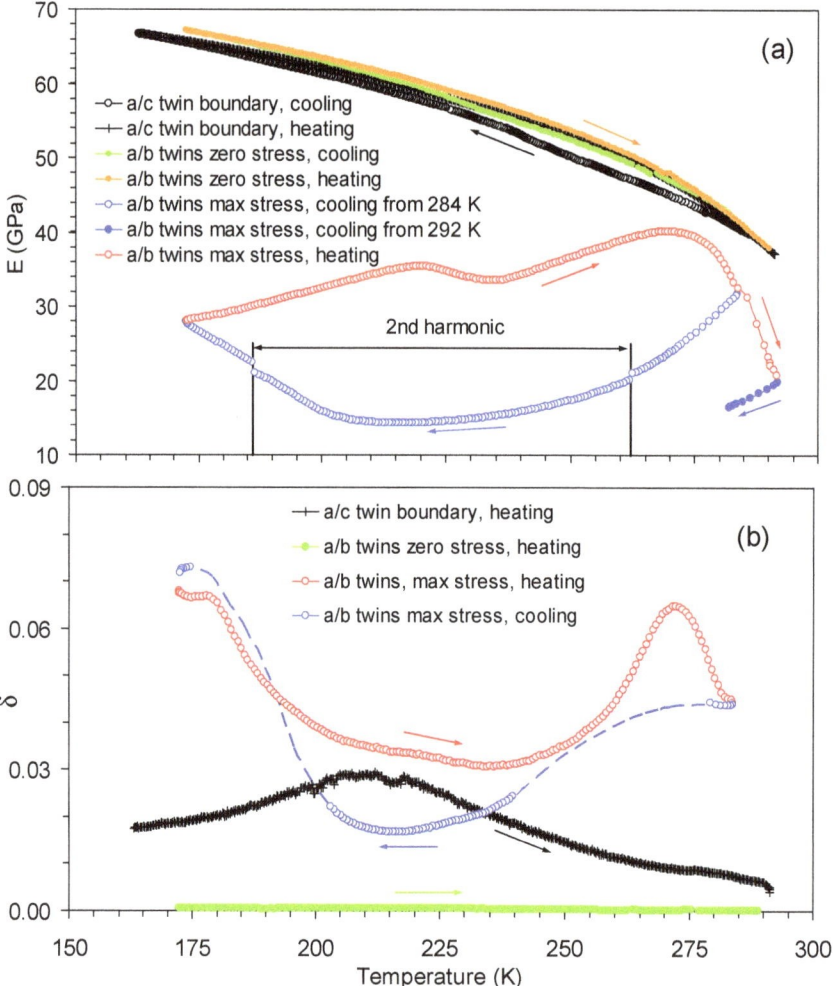

Figure 9. Temperature spectra of the Young's modulus (**a**) and internal friction (**b**) for the samples with three types of *a/c* and *a/b* twin structures shown in Figure 1. Cooling and heating scans at 2K/min are shown in (**a**) and in (**b**) for the Structure 3 sample with maximum shear stress in the *a/b* twinning planes. For clarity, only heating scans are shown in (**b**) for the Structure 1 sample with *a/c* twin boundaries and Structure 2 sample (zero *a/b* shear stress), for which the thermal hysteresis is not significant. Note that the data for the Structures 2 and 3 (with zero and maximum *a/b* shear stress) correspond to the same sample with *a/b* twin structure reoriented by magnetic field, see [33] for the details of *a/b* twin structure modification. Broken parts of the temperature spectrum on cooling for the sample Structure 3 (maximum shear stress for *a/b* twins) mark the ranges where the damping values cannot be determined correctly due to the change of the oscillation harmonic (see text for details).

These anomalies and difficulty in defining the YM have previously been attributed to the extremely high a/b TB-related contribution of the anelastic strain to the apparent YM of the a/b laminate [33]. All abovementioned features of the YM behaviour point to a highly non-equilibrium state of the a/b twin structure over a broad temperature range. Probably, the a/b twin-related dramatic variations of the YM reflect recently reported reversible transitions to the nanotwin structure upon cooling of 10 M martensite [22].

Figure 9a shows that the YM temperature spectrum for the sample with the central a/c twin (Structure 1) has certain (minor) features of the behaviour of the Structure 3 sample with dominating role of a/b twins. Namely, the modulus for the Structure 1 sample is slightly lower than for the Structure 2 sample without the central narrow variant. Second, a slight thermal hysteresis can be discerned in the YM spectrum for the Structure 1 sample. These minor details should be attributed to the contribution of a/b twins in the narrow central variant with c-axis across the sample axis.

The data on damping for the Structure 3 sample, Figure 9b, deserve a brief comment. The data on heating scans are perfectly reliable. However, during cooling, due to the strong variations of the YM, the oscillation mode of the sample switches from the fundamental to the second harmonic. The second harmonic is maintained over a broad temperature range marked in Figure 9a. Switching between the fundamental and second harmonic has only a minor effect on the derived YM value, seen as small steps during the change of the oscillation mode, Figure 9a. In contrast to this minor effect in YM, the change of the harmonics very strongly affects the IF spectra on cooling. More specifically, it provokes intense IF peaks due to the strong displacement of the stress node from the junction between the sample and quartz transducers. Therefore, the IF of the material cannot be registered reliably over the ranges of the harmonics change. Orientative behaviour of the IF over these ranges is indicated in Figure 9b by a discontinuous line. Nevertheless, the temperature of the transitory IF peak, 215 K, is well away from the harmonics switching points and Figure 9b shows that the in the Structure 3 sample transient IF peak at around 215 K does not exist on cooling and heating.

It is important to emphasize that in the Structure 3 sample with a/b twins under maximum stress, in addition to the absence of the transitory damping peak, T-dot transient effects are not detected either over the entire temperature range studied, despite the intense modification of the a/b twin structure during thermal cycling evidenced by very intense apparent modulus variations. Thus, the transitory IF is inherent only in the sample containing a/c TBs. This is a strong argument in favour of the magnetic nature of the IF transitory peak and T-dot effect: motion of a/b TBs produces additional strain affecting YM, but does not produce magnetic flux variations since the orientation of the easy axis remains unchanged. On the contrary, any displacement of the a/c TB results in the local magnetic flux variation and, hence, eddy current-related IF of magnetic origin.

In summary, the study of samples with different twin structures indicates:
- thermal hysteresis of the IF and YM in 10 M martensite is exclusively related to the a/b twins and reflects their non-equilibrium state; the hysteresis is practically absent in the sample with a/c TBs and when shear stress in a/b twinning planes does not exist;
- a/c TB is responsible for the transitory IF peak; the peak is absent in samples without a/c TBs for both orientations of a/b twins.

4.3. Origin of T-Dot and Time Dependence of the Internal Friction

A simultaneous occurrence of both time and T-dot dependence of the IF during interruption of temperature scan has been found and analysed in the case of antiferromagnetic Dy [13]. In the present results, the former should be associated with pinning of TBs by mobile obstacles detected in strain amplitude scans, Figure 3, the latter—with the motion of the a/c TBs under thermal stresses during temperature variations.

As for the T-dot dependence, we mention first that a temperature-induced displacement of even relatively low-mobility Type I TB can be detected in the results presented by Straka et al. [22]. In the optical images of Figure 5 [22], the position of the Type I TB changes with temperature. This displacement is seen for temperatures 298 and 223 K, Figure 5a,d and becomes obvious if the position of the twin boundary at 223 K is compared with the trace of this boundary as seen in the austenite at 333 K, cf. Figure 5d,f. The TB displacement can be evaluated in Figure 5 of Ref. [22] as around only 0.1 µm. However, the TB motion must be much more intense in our "domain engineered" sample for the following two reasons:

- the overall twin structure of the sample has not been identified in Ref. [22], which was supposedly polyvariant; so TB interaction is expected to substantially inhibit their mutual displacement; in other words, a single TB must be much more mobile than TBs of the polyvariant structure;
- twin boundary shown in Figure 5 of Ref. [22] is of Type I type, much less mobile than Type II TB studied in the present work.

Observations of the T-dot effect in the Structure 1 sample with a/c TB and its absence in the Structure 3 sample, despite intense rearrangement of the a/b twin structure evidenced by the YM spectra, confirms our conclusion on the magnetic origin of the uncovered ultrasonic T-dot effect. Indeed, the displacement of the a/c TB provokes local magnetic flux change due to the reorientation of the easy axis of magnetization over the volume swept by moving TB, whereas the displacement of a/b TBs produces only strain, without c-axis reorientation and magnetic flux change. The rearrangement of magnetization of parts of magnetic domains adjacent to a/c TB during the motion of the boundary has been directly observed in a $Ni_{53.8}Mn_{23.7}Ga_{22.5}$ alloy [44] by means of high-resolution Interference–Contrast–Colloid method.

A number of effects observed in the present study is perfectly consistent with the concept of TBs moving quasistatically under thermal stresses in the presence of mobile obstacles. For a detailed analysis and examples of similar cases in Cu-Al-Ni, Cu-Al-Be and Ni-Fe-Ga shape memory alloys we refer to a review [37]. Strain amplitude hysteresis and jerky IF dependences on strain amplitude, Figure 3, point to the intense pinning of a/c TBs by mobile obstacles at room temperature. The quasistatic motion of TBs at elevated temperatures (e.g., room temperature) is accompanied by the so-called "dragging" [45] of pinning clouds. Upon lowering the temperature, the mode of this motion changes from the viscous-like (which is actually a sequence of small macroscopically undistinguishable jerks) to the intermittent and macroscopically jerky one due to a competition between dragging and depinning mechanisms. Eventually, at low enough temperatures, even the TB-assisted diffusion of point-like defects freezes out. The TBs leave behind the Cottrell-like clouds and move in a random potential of homogeneously distributed immobile obstacles. The density of Cottrell-like clouds during motion of TBs is lower than under isothermal conditions and the IF decrease upon interruptions of temperature scans corresponds to the increase in the density of clouds or time-dependent pinning of TBs. Figure 7b shows that the recuperation of TB motion upon resuming the temperature scan requires overcooling/overheating and occurs as a fast depinning or jerk.

5. Conclusions

In general terms of anelasticity of solids, present experiments evidence the existence of the ultrasonic transitory internal friction of magnetic origin. This type of internal friction can be classified as a new category that links the classes of transitory and magnetomechanical damping.

In terms of the structure and properties of 10 M Ni-Mn-Ga martensite, the use of "domain engineered" samples with different controlled twin configurations shed light on the a/c and a/b twin boundary dynamics. More specifically, we conclude that:

- a/c twin boundaries are pinned at and somewhat below room temperature by Cottrell-like clouds of mobile obstacles, whereas a/b twin boundaries are free from pinning clouds;
- the anisotropy of thermal expansion results in the motion of a/c twin boundaries during temperature variations; this displacement provokes, on one hand, local changes

of the magnetic flux and T-dot ultrasonic transitory internal friction of magnetic origin; on the other hand, the quasistatic motion of a/c twin boundaries under thermal stresses occurs under conditions similar to dynamic strain ageing, manifested in the time dependence of the internal friction due to the interaction of twin boundaries with Cottrell-like clouds of more or less mobile (depending on the temperature) pinners;
- over a broad temperature range the system of a/b twins is in a highly non-equilibrium state, resulting in a strong history dependence and dramatic variations of the apparent Young's modulus with temperature.

Author Contributions: Conceptualization, S.K., A.S. and K.U.; methodology, S.K. and A.S.; investigation, S.K., A.S., B.D. and K.S.; writing—original draft preparation, S.K.; writing—review and editing, A.S., B.D., K.S. and V.N.; supervision, K.U. All authors have read and agreed to the published version of the manuscript.

Funding: This work was supported by the Spanish AEI and FEDER, UE (project RTI 2018-094683-BC51), Strategic Research Council of Finland (grant number 313349) and the Academy of Finland (grant number 325910).

Acknowledgments: The authors are grateful to A. Sozinov (LUT) for supplying the material and to R. Sánchez Torres (SCT-UIB) for the help in preparation of experiments.

Conflicts of Interest: The authors declare no conflict of interest.

References

1. San Juan, J.; Pérez-Saez, R.B. 5.4 Transitory Effects. *Mater. Sci. Forum* **2001**, *366–368*, 416–436. [CrossRef]
2. Lebedev, A.B. Internal Friction in Quasi-Static Crystal Deformation (Review). *Sov. Phys. Solid State* **1993**, *35*, 2305–2341.
3. Kustov, S.; Golyandin, S.; Sapozhnikov, K.; Vincent, A.; Maire, E.; Lormand, G. Structural and Transient Internal Friction Due to Thermal Expansion Mismatch between Matrix and Reinforcement in Al–SiC Particulate Composite. *Mater. Sci. Eng. A* **2001**, *313*, 218–226. [CrossRef]
4. Sapozhnikov, K.; Golyandin, S.; Kustov, S. Elastic and Anelastic Properties of C/Mg-2wt.%Si Composite during Thermal Cycling. *Compos. Part A* **2009**, *40*, 105–113. [CrossRef]
5. L'vov, V.A.; Glavatska, N.; Aaltio, I.; Söderberg, O.; Glavatskyy, I.; Hannula, S.P. The Role of Anisotropic Thermal Expansion of Shape Memory Alloys in Their Functional Properties. *Acta Mater.* **2009**, *57*, 5605–5612. [CrossRef]
6. Sapozhnikov, K.; Golyandin, S.; Kustov, S.; Van Humbeeck, J.; Schaller, R.; De Batist, R. Transient Internal Friction during Thermal Cycling of Cu-Al-Ni Single Crystals in $\beta_{1'}$ Martensitic Phase. *Scr. Mater.* **2002**, *47*, 459–465. [CrossRef]
7. Sapozhnikov, K.; Golyandin, S.; Kustov, S.; Schaller, R.; Van Humbeeck, J. Anelasticity of B19' Martensitic Phase in Ni-Ti and Ni-Ti-Cu Alloys. *Mater. Sci. Eng. A* **2006**, *442*, 398–403. [CrossRef]
8. Coronel, V.F.; Beshers, D.N. Magnetomechanical Damping in Iron. *J. Appl. Phys.* **1988**, *64*, 2006–2015. [CrossRef]
9. Degauque, J. Magnetic Domains. In *Mechanical Spectroscopy Q-1 2001 with Applications to Materials Science*; Schaller, R., Fantozzi, G., Gremaud, G., Eds.; Trans Tech Publications: Uetikon, Zuerich, 2001; pp. 453–482.
10. Kustov, S.; Corró, M.L.; Kaminskii, V.; Saren, A.; Sozinov, A.; Ullakko, K. Elastic and Anelastic Phenomena Related to Eddy Currents in Cubic Ni$_2$MnGa. *Scr. Mater.* **2018**, *147*, 69–73. [CrossRef]
11. Birchak, J.R.; Smith, G.W. Magnetomechanical Damping and Magnetic Properties of Iron Alloys. *J. Appl. Phys.* **1972**, *43*, 1238–1246. [CrossRef]
12. Mason, W.P. Domain Wall Relaxation in Nickel. *Phys. Rev.* **1951**, *83*, 683–684. [CrossRef]
13. Kustov, S.; Liubimova, I.; Corró, M.; Torrens-Serra, J.; Wang, X.; Haines, C.R.S.; Salje, E.K.H. Temperature Chaos, Memory Effect, and Domain Fluctuations in the Spiral Antiferromagnet Dy. *Sci. Rep.* **2019**, *9*, 5076. [CrossRef]
14. Transitory damping stems from the superposition of oscillatory and translatory motion of magnetic domain walls. A theoretical solution for the magnetic transitory damping yields a superposition of two terms: Frequency-independent and the one that shows conventional inverse frequency dependence, to be published.
15. Likhachev, A.A.; Sozinov, A.; Ullakko, K. Different Modeling Concepts of Magnetic Shape Memory and Their Comparison with Some Experimental Results Obtained in Ni-Mn-Ga. *Mater. Sci. Eng. A* **2004**, *378*, 513–518. [CrossRef]
16. Fukuda, T.; Sakamoto, T.; Kakeshita, T.; Takeuchi, T.; Kishio, K. Rearrangement of Martensite Variants in Iron-Based Ferromagnetic Shape Memory Alloys under Magnetic Field. *Mater. Trans.* **2004**, *45*, 188–192. [CrossRef]
17. Chernenko, V.A.; L'vov, V.A. Magnetoelastic Nature of Ferromagnetic Shape Memory Effect. *Mater. Sci. Forum* **2008**, *583*, 1–20. [CrossRef]
18. Rolfs, K.; Mecklenburg, A.; Guldbakke, J.M.; Wimpory, R.C.; Raatz, A.; Hesselbach, J.; Schneider, R. Crystal Quality Boosts Responsiveness of Magnetic Shape Memory Single Crystals. *J. Mag. Mag. Mater.* **2009**, *321*, 1063–1067. [CrossRef]

19. Straka, L.; Lanska, N.; Ullakko, K.; Sozinov, A. Twin Microstructure Dependent Mechanical Response in Ni-Mn-Ga Single Crystals. *Appl. Phys. Lett.* **2010**, *96*, 131903. [CrossRef]
20. Sozinov, A.; Lanska, N.; Soroka, A.; Straka, L. Highly Mobile Type II Twin Boundary in Ni-Mn-Ga Five-Layered Martensite. *Appl. Phys. Lett.* **2011**, *99*, 124103. [CrossRef]
21. Seiner, H.; Straka, L.; Heczko, O. A microstructural model of motion of macro-twin interfaces in Ni–Mn–Ga 10M martensite. *J. Mech. Phys. Sol.* **2014**, *64*, 198–211. [CrossRef]
22. Straka, L.; Drahokoupil, J.; Veřtát, P.; Zelený, M.; Kopeček, J.; Sozinov, A.; Heczko, O. Low Temperature *a/b* Nanotwins in $Ni_{50}Mn_{25+x}Ga_{25-x}$ Heusler Alloys. *Sci. Rep.* **2018**, *8*, 11943. [CrossRef]
23. Ullakko, K.; Huang, J.K.; Kantner, C.; O'Handley, R.C.; Kokorin, V.V. Large Magnetic-field-induced Strains in Ni_2MnGa Single Crystals. *Appl. Phys. Lett.* **1996**, *69*, 1966–1968. [CrossRef]
24. Aaltio, I.; Sozinov, A.; Ge, Y.; Ullakko, K.; Lindroos, V.K.; Hannula, S.-P. *Reference Module in Materials Science and Materials Engineering*; Hashmi, S., Ed.; Elsevier: Oxford, UK, 2016; pp. 1–14.
25. Lanska, N.; Söderberg, O.; Sozinov, A.; Ge, Y.; Ullakko, K.; Lindroos, V.K. Composition and Temperature Dependence of the Crystal Structure of Ni-Mn-Ga Alloys. *J. Appl. Phys.* **2004**, *95*, 8074–8078. [CrossRef]
26. Chulist, R.; Straka, L.; Lanska, N.; Soroka, A.; Sozinov, A.; Skrotzki, W. Characterization of Mobile Type i and Type II Twin Boundaries in 10M Modulated Ni-Mn-Ga Martensite by Electron Backscatter Diffraction. *Acta Mater.* **2013**, *61*, 1913–1920. [CrossRef]
27. Heczko, O.; Klimša, L.; Kopeček, J. Direct Observation of A-b Twin Laminate in Monoclinic Five-Layered Martensite of Ni-Mn-Ga Magnetic Shape Memory Single Crystal. *Scr. Mater.* **2017**, *131*, 76–79. [CrossRef]
28. Saren, A.; Sozinov, A.; Kustov, S.; Ullakko, K. Stress-Induced a/b Compound Twins Redistribution in 10M Ni-Mn-Ga Martensite. *Scr. Mater.* **2020**, *175*, 11–15. [CrossRef]
29. Robinson, W.H.; Edgar, A. The Piezoelectric Method of Determining Mechanical Damping at Frequencies of 30 to 200 KHz. *IEEE Trans. Sonics Ultrason.* **1974**, *21*, 98–105. [CrossRef]
30. Kustov, S.; Golyandin, S.; Ichino, A.; Gremaud, G. A New Design of Automated Piezoelectric Composite Oscillator Technique. *Mater. Sci. Eng. A* **2006**, *442*, 532–537. [CrossRef]
31. Read, T.A. The Internal Friction of Single Metal Crystals. *Phys. Rev.* **1940**, *58*, 371–380. [CrossRef]
32. Salje, E.K.H. Multiferroic Domain Boundaries as Active Memory Devices: Trajectories towards Domain Boundary Engineering. *Chem. Phys. Chem.* **2010**, *11*, 940–950. [CrossRef]
33. Kustov, S.; Saren, A.; Sozinov, A.; Kaminskii, V.; Ullakko, K. Ultrahigh Damping and Young's Modulus Softening Due to a/b Twins in 10M Ni-Mn-Ga Martensite. *Scr. Mater.* **2020**, *178*, 483–488. [CrossRef]
34. Kustov, S.; Liubimova, I.; Salje, E.K.H. LaAlO3: A Substrate Material with Unusual Ferroelastic Properties. *Appl. Phys. Lett.* **2018**, *112*, 042902. [CrossRef]
35. Kustov, S.; Golyandin, S.; Sapozhnikov, K.; Van Humbeeck, J.; de Batist, R. Low-Temperature Anomalies in Young's Modulus and Internal Friction of Cu–Al–Ni Single Crystals. *Acta Mater.* **1998**, *46*, 5117–5126. [CrossRef]
36. Sapozhnikov, K.; Golyandin, S.; Kustov, S.; Van Humbeeck, J.; de Batist, R. Motion of Dislocations and Interfaces during Deformation of Martensitic Cu–Al–Ni Crystals. *Acta Mater.* **2000**, *48*, 1141–1151. [CrossRef]
37. Kustov, S.; Sapozhnikov, K.; Wang, X. Phenomena Associated with Diffusion, Assisted by Moving Interfaces in Shape Memory Alloys. A review of our Previous Works. *Funct. Mater. Lett.* **2017**, *10*, 1740010. [CrossRef]
38. Chernenko, V.A.; Seguí, C.; Cesari, E.; Pons, J.; Kokorin, V.V. Sequence of Martensitic Transformations in Ni-Mn-Ga Alloys. *Phys. Rev. B* **1998**, *57*, 2659–2662. [CrossRef]
39. Liu, J.; Wang, J.; Jiang, C.; Xu, H. Internal Friction Associated with the Premartensitic Transformation and Twin Boundary Motion of $Ni_{50+x}Mn_{25-x}Ga_{25}$ (x=0–2) Alloys. *J. Appl. Phys.* **2013**, *113*, 103502. [CrossRef]
40. Chang, S.H.; Wu, S.K. Low-Frequency Damping Properties of near-Stoichiometric Ni_2MnGa Shape Memory Alloys under Isothermal Conditions. *Scr. Mater.* **2008**, *59*, 1039–1042. [CrossRef]
41. Wang, W.H.; Liu, G.D.; Wu, G.H. Magnetically Controlled High Damping in Ferromagnetic $Ni_{52}Mn_{24}Ga_{24}$ Single Crystal. *Appl. Phys. Lett.* **2006**, *89*, 101911. [CrossRef]
42. Wang, W.H.; Ren, X.; Wu, G.H. Martensitic Microstructure and Its Damping Behavior in $Ni_{52}Mn_{16}Fe_8Ga_{24}$ Single Crystals. *Phys. Rev. B* **2006**, *73*, 092101. [CrossRef]
43. Gavriljuk, V.G.; Söderberg, O.; Bliznuk, V.V.; Glavatska, N.I.; Lindroos, V.K. Martensitic transformations and mobility of twin boundaries in Ni_2MnGa alloys studied by using internal friction. *Scr. Mater.* **2003**, *49*, 803–809. [CrossRef]
44. Chopra, H.D.; Ji, C.; Kokorin, V.V. Magnetic-Field-Induced Twin Boundary Motion in Magnetic Shape-Memory Alloys. *Phys. Rev. B* **2000**, *61*, R14913–R14915. [CrossRef]
45. Gremaud, G. Dislocation-point defect interactions. In *Mechanical Spectroscopy Q-1 2001 with Applications to Materials Science*; Schaller, R., Fantozzi, G., Gremaud, G., Eds.; Trans Tech Publications: Ueticon, Zuerich, 2001; pp. 178–246.

Article

A Ternary Map of Ni–Mn–Ga Heusler Alloys from Ab Initio Calculations

Yulia Sokolovskaya [1], Olga Miroshkina [1,2], Danil Baigutlin [1,3], Vladimir Sokolovskiy [1,4,*], Mikhail Zagrebin [1,4,5], Vasilly Buchelnikov [1,4,*] and Alexey T. Zayak [6]

[1] Condensed Matter Physics Department, Chelyabinsk State University, 454001 Chelyabinsk, Russia; sya2890@mail.ru (Y.S.); miroshkina.on@yandex.ru (O.M.); d0nik1996@mail.ru (D.B.); miczag@mail.ru (M.Z.)
[2] Faculty of Physics and Center for Nanointegration Duisburg-Essen (CENIDE), University of Duisburg-Essen, 47048 Duisburg, Germany
[3] Department of Physics, School of Engineering Science, LUT University, FI-53850 Lappeenranta, Finland
[4] Academic Research Center for Energy Efficiency, National University of Science and Technology "MISiS", 119049 Moscow, Russia
[5] Faculty of Mathematics, Mechanics and Computer Sciences, South Ural State University (National Research University), 454080 Chelyabinsk, Russia
[6] Department of Physics and Astronomy, Bowling Green State University, Bowling Green, OH 43403, USA; azayak@bgsu.edu
* Correspondence: vsokolovsky84@mail.ru (V.S.); buche@csu.ru (V.B.); Tel.: +7-35-1799-7117 (V.S.)

Citation: Sokolovskaya, Y.; Miroshkina, O.; Baigutlin, D.; Sokolovskiy, V.; Zagrebin, M.; Buchelnikov, V.; Zayak, A.T. A Ternary Map of Ni–Mn–Ga Heusler Alloys from Ab Initio Calculations. *Metals* **2021**, *11*, 973. https://doi.org/10.3390/met11060973

Academic Editor: Tadeusz Kulik

Received: 12 May 2021
Accepted: 13 June 2021
Published: 17 June 2021

Publisher's Note: MDPI stays neutral with regard to jurisdictional claims in published maps and institutional affiliations.

Copyright: © 2021 by the authors. Licensee MDPI, Basel, Switzerland. This article is an open access article distributed under the terms and conditions of the Creative Commons Attribution (CC BY) license (https://creativecommons.org/licenses/by/4.0/).

Abstract: In the search for new magnetic functional materials, non-stoichiometric compounds remain a relatively unexplored territory. While experimentalists create new compositions looking for improved functional properties, their work is not guided by systematic theoretical predictions. Being designed for perfect periodic crystals, the majority of first-principles approaches struggle with the concept of a non-stoichiometric system. In this work, we attempt a systematic computational study of magnetic and structural properties of Ni–Mn–Ga, mapped onto ternary composition diagrams. Compositional stability was examined using the convex hull analysis. We show that the cubic austenite has its stability region close to the stoichiometric Ni_2MnGa, in agreement with experimental data, while the tetragonal martensite spreads its stability over a wider range of Mn and Ni contents. The unstable compositions in both austenite and martensite states are located in the Ga-rich corner of the ternary diagram. We note that simultaneous stability of the austenite and martensite should be considered for potentially stable compounds suitable for synthesis. The majority of compounds are predicted to be ferrimagnetically ordered in both austenitic and martensitic states. The methodology used in this work is computationally tractable, yet it delivers some predictive power. For experimentalists who plan to synthesize stable Ni–Mn–Ga compounds with ferromagnetic order, we narrow the target compositional range substantially.

Keywords: Heusler alloys; ternary diagrams; structural phase stability; first-principles approach

1. Introduction

During the last decades, shape memory Ni–Mn-based Heusler alloys have received notable attention as promising materials for a wide variety of engineering applications such as a magnetic actuator, controller, sensor, and damping technologies. Ni_2MnGa is probably the most known Heusler alloy. Its distinguishing feature is the martensitic transformation in the ferromagnetic (FM) state between the high-temperature cubic austenite and the low-temperature modulated martensite with twin boundary structure occurring at about $T_m \approx 200$ K [1]. In 1996, Ulakko et al. [2] suggested the idea of magnetically induced reorientation of the twin variants structure in Ni_2MnGa. These authors were the first to demonstrate 0.2% field-induced strains along the (001) direction in a Ni_2MnGa single crystal with a magnetic field of 0.8 T applied at 265 K. That success was followed by extensive research of non-stoichiometric Ni–Mn–Ga compositions, covering various aspects such as the crystal structure of the austenite and martensite [3–7], magnetic and magnetoresistance properties [8,9], thermally and magnetically induced deformation [2,10,11], magnetocaloric properties [12,13], heat treatment processes [3,14], and phase diagrams [15–22].

At present, the most studied off-stoichiometric compounds are concentrated near Ni_2MnGa. However, it is clear that expanding that region can lead to new functional phenomena. For example, we know that the martensitic and magnetic transition temperatures are sensitive to the valence electron concentration e/a (electrons per atom). Ni-rich $Ni_{2+x}Mn_{1-x}Ga$ ($0 \leq x < 0.4$ and $7.5 \leq e/a < 7.8$) [18] and Mn-rich $Ni_2Mn_{1+x}Ga_{1-x}$ ($0 \leq x < 0.6$ and $7.5 \leq e/a < 8.0$) [22] provide an example. Some of these compositions reveal a strong magnetostructural coupling leading to an appearance of the first-order magnetostructural phase transition between the FM or antiferromagnetic (AFM) (ferrimagnetic (FIM)) martensite and the FM or paramagnetic (PM) austenite.

Compositions close to stoichiometric Mn_2NiGa and Ga_2MnNi are not investigated so deeply. According to experimental findings [23–25], Mn_2NiGa austenite possesses an inverse Heusler structure ($F\bar{4}3m$, space group no. 216) and undergoes a martensitic transformation in the FIM state at ≈ 270 K. In addition, a large thermal hysteresis of 50 K and a high Curie temperature $T_C = 588$ K are reported. Ab initio studies Reference [26–28] confirm the inverse atomic arrangement. Ga_2MnNi is found to be an FM martensite at room temperature with T_C of about 330 K. The martensitic transformation in the PM state takes place at $\approx$780 K, and this temperature is the highest reported so far in the Ni–Mn–Ga family [29]. Experimental and theoretical study [29] revealed the regular $L2_1$ structure of the austenitic phase, in which Ga atoms occupy equivalent positions.

Overall, we see that properly chosen deviations in stoichiometry can tune the magnetostructural coupling in Ni–Mn–Ga; however, the data are still sparse. Experimental studies of non-stoichiometric systems are associated with the complexity of synthesis as well as challenges of mechanical and segregational instabilities. Meaningful and systematic predictions from first-principles computations could allow for more efficient strategies of exploring wide ranges of non-stoichiometric compositions. Earlier, ab initio methods were used to explore structural, magnetic, vibrational, and thermodynamic properties of $Ni_{2+x}Mn_{1-x}Ga$ and $Ni_2Mn_{1+x}Ga_{1-x}$ alloys (for instance, see References [30–36]).

It was demonstrated that doping of Ni–Mn–Ga significantly influences lattice constants, bulk moduli, magnetic moments, and Curie temperatures of the austenite and provided qualitative information about the phase stability in terms of the formation energy [37–39].

Along with the conventional first-principles methods, chemical disorder in $Ni_xMn_yGa_z$ was simulated using the coherent potential approximation (CPA) [37,38,40]. While this method treats atomic disorder very efficiently, it lacks the ability for complete structural relaxations, leaving some uncertainty that relies on the average description of the calculated properties of a crystal with ideal atomic positions. In order to verify our CPA results, we utilized simulations of the atomic disorder in Ni–Mn–Ga within 16-atom supercells, in [39], using the projector augmented wave (PAW) method [41,42]. Both approaches revealed similar compositional trends, which makes sense, since we did not allow the PAW calculations to fully relax atomic positions, keeping all atoms in the high-symmetry sites. Additional studies [37–39] showed similar compositional trends.

Thus, one significant drawback of our previous studies was the lack or internal relaxations of atomic positions, i.e., we assumed that in the high-symmetry cubic phase, atoms would naturally stick to the high-symmetry sites. Moreover, our previous work was focused on the stability of the austenitic phase only, assuming that martensite would have the same stability range. However, further analysis shows that the prediction of stable FM compositions requires simultaneous stability of the austenite and martensite. The challenge with this approach is that martensite requires relaxation of all degrees of freedom, allowing atoms to leave the high-symmetry sites. For Ni–Mn–Ga, local relaxations are essential, leading to such phenomena as structural modulations, which influences lattice parameters, structural stability, and magnetic order. Optimizing all those degrees of freedom for all possible compositions is a formidable task; therefore, we have to develop a simplified approach, which would be computationally manageable, and make a step in the right direction toward reliable ab initio predictions of compositional trends.

In this work, we analyze compositional trends in austenitic and martensitic phases of $Ni_xMn_yGa_z$ alloys over the full ternary composition ranges. Our analysis of the chemical stability utilizes the convex hull methodology, which takes into account most possible reactions, providing a direct measure of the phase stability from the first-principles study [43].

2. Computational Methodology

Magnetic and structural properties were calculated using the PAW method implemented in VASP [41,42]. The exchange correlation effects were treated with the Perdew–Burke–Ernzerhof functional of the generalized gradient approximation [44]. The PAW pseudopotentials with the following electronic configurations were used: $3p^63d^84s^2$ for Ni, $3p^63d^64s^1$ for Mn, and $3d^{10}4s^24p^1$ for Ga. A uniform Monkhorst-Pack mesh of $8 \times 8 \times 8$ k-points was used to sample the Brillouin zone with the first-order Methfessel–Paxton method ($\sigma = 0.1$ eV). The plane-wave basis kinetic-energy cutoff of 400 eV was applied, whereas the kinetic-energy cutoff for the augmented charge was chosen as 800 eV. Full crystal structure optimization, including atomic coordinates and lattice parameters, was performed using the conjugate-gradient algorithm. The convergence criterion for the residual forces is 10^{-5} eV/Å, while the self-consistent calculations are converged with an accuracy of 10^{-6} eV.

We used 16-atom supercells to model 105 compositions covering uniformly the ternary Ni–Mn–Ga diagram, as shown in Figure 1. In order to facilitate discussions, we split the diagram schematically into the three zones depending on Mn content y and denoted them as Area I ($y \leq 25$ at.%), II ($25 < y < 50$ at.%), and III ($y \geq 50$ at.%).

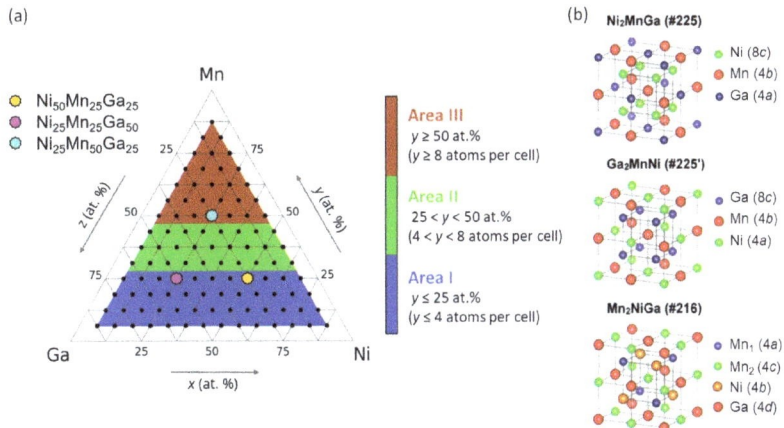

Figure 1. (Color online) (**a**) The set of 105 composition points formed by the 16-atom supercells and represented as a ternary diagram of the Ni–Mn–Ga system. Color circles denote the stoichiometric compositions. (**b**) Regular (#225 and #225′) and inverse (#216) cubic Heusler structures with the corresponding Wyckoff positions for the stoichiometric compounds.

The choice of these areas based on Mn content is justified by similarities in magnetic configurations for each area, which are attributed to Mn atoms sitting in different sublattices. We examine both regular ($Fm\bar{3}m$, space group #225) and inverse ($F\bar{4}3m$, space group no. 216) Heusler structures for all compositions in the austenitic phase and find that the regular structure is preferable for Ni- and Ga-rich compositions (areas I and II). In this case, the alloys assume the #225 and #225′ structures, as shown in Figure 1. The inverse Heusler structure turns out to be more favorable for Mn-rich compositions (area III).

One can find more information about crystal structures in the Supplementary Material (SM) (Table S1 and Figures S1 and S2).

The magnetic ordering of Mn-excess compositions is rather complex due to the magnetic moments of Mn atoms that occupy the Ga and Ni sites and can interact antiferromagnetically with those on the Mn sublattice. In view of this fact, we considered several types of magnetic configurations, which can be grouped as follows: (i) area I: FM only; (ii) area II: FM, FIM-1, and FIM-2 (Mn excess atoms have a reversed magnetic moment to that of Mn atoms on the Mn sublattice); (iii) area III: FM one and seven FIM configurations. All magnetic configurations are listed in Table S2 and visualized in Figures S3 and S4 of the SM.

To analyze the phase stability of the compounds, we used a new three-step procedure, with an improvement over our previous studies [37–39], where only the formation energy (E_{form}) was used as the stability criterion. At first, we evaluate the stability of austenitic (martensitic) phases in the context of the preservation of cubic (tetragonal/orthorhombic) symmetry after the full relaxation for each composition. In the second step, the E_{form} of the compounds in austenitic and martensitic phases is analyzed. It is assumed that a compound is stable against the decomposition into its pure components if E_{form} has a negative sign in accordance with the equation $E_{form} = E_0^{Ni_xMn_yGa_z} - (xE_0^{Ni} + yE_0^{Mn} + zE_0^{Ga})$. Here $E_0^{Ni_xMn_yGa_z}$ is the ground-state energy of $Ni_xMn_yGa_z$, while E_0^{Ni}, E_0^{Mn}, and E_0^{Ga} are the ground-state energies of corresponding components in their preferable structures. Compositions with a positive E_{form} are eliminated from the consideration. In the third step, the phase stability is determined using the convex hull analysis in conjunction with the evaluation of decomposition into pure elements and stable binary compounds. To plot the convex hull, 13 pivot points were used as follows: Ga, Ni, Mn, Ga_3Ni_2, Ga_3Ni_5, Ga_5Ni, Ga_7Ni_3, Ga_9Ni_{13}, $GaNi_3$, $MnGa$, $MnGa_4$, $MnNi_3$, and Ni_2MnGa. Structural parameters for the pivot points are listed in SM (Table S3). To investigate the decomposition trends, all possible combinations of the reaction products (pivot points) were generated for each compound. We took into account the decomposition of two and three reaction products, each of these might be one of 13 pivot points. The stoichiometric coefficients for reactions are evaluated by solving linear equations, only considering equations with the positive stoichiometric coefficients. For each composition, a fraction of stable reactions is normalized on its own number of possible reactions depending on Ni, Mn, and Ga content. By an example of $Ni_9Mn_4Ga_3$, 27 possible decomposition reactions are listed in Table S4 of the SM. This compound is stable against decomposition in the case of 22 reactions, which gives us 81% of stable reactions.

3. Results and Discussion

To gain information about the structural stability of compounds under study, we first discuss the arrangement of atoms within a crystalline cubic structure for each composition. Figure 2 illustrates the ternary diagram of the cubic structures distribution with the preferable atomic arrangement for $Ni_xMn_yGa_z$.

The full geometric optimization of structures with the cubic supercell suggested that 54 out of the 105 compositions preserve a cubic symmetry, as opposed to Reference [39] where all 105 compositions were considered with the cubic structure. Another distinctive feature of the full geometric optimization (as opposed to the relaxation without change in atomic positions) is the prediction of slightly different atomic configurations. Thus, the favorable configurations of atoms are turned out to be as follows: (i) the Mn-rich compounds ($y \geq 50$ at.%) have the inverse structure (#216); (ii) the compounds of $x > 45$ at.%Ni, $25 < y < 50$ at.%Mn, and $z < 30$ at.%Ga possess the regular structure (#225) in which Ni atoms occupy $8c$ sites; (iii) the compounds of $x < 45$ at.%Ni and $y < 25$ at.%Mn tend to favor the #225' structure, in which Ga atoms are located at $8c$ sites. However, neglecting of the atomic relaxations [39] does not provide the #225^{mix} atomic ordering for compositions with $30 < x < 45$ at.%Ni, $6 < y < 25$ at.%Mn, and $40 < z < 70$ at.%Ga and the #216 one for $Mn_{50}Ni_{25}Ga_{25}$. In the latter case, the inverse Heusler structure (#216) has been observed experimentally [26,27]. For the remaining 51 compounds, the initial cubic structure is transformed into the tetragonal and orthorhombic configurations. Among those are 27 compounds mainly with Ga- and Ni-rich content from area I; 15 compounds of area II with variable content of the all constituent elements; and 9 Mn-rich compositions with $z < 15$ at.%Ga from the area III.

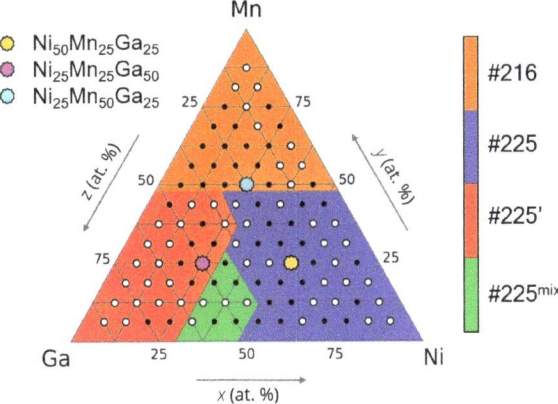

Figure 2. (Color online) Distribution of stable austenitic crystal structures mapped into the ternary diagram of Ni$_x$Mn$_y$Ga$_z$. Open circles indicate the compositions with tetragonally and orthorhombically distorted structures afforded by the geometric optimization procedure of an initial cubic structure. Here #225^{mix} indicates the regular structure with 8c sites simultaneously occupied by Ni and Ga.

At the same time, a full geometric optimization of structures with a tetragonal supercell reveals 13 compositions, in which the tetragonal structures are higher in energy than the cubic one. These are 11 Ga-rich and near-equiatomic Ni and Ga compositions from the area I and two compositions with nearly equal Mn and Ga content from the area II. The list of compositions with stable austenite and martensite phases is presented in Tables S5–S10, see the SM.

Figure 3 shows formation energies for compounds in the austenitic and martensitic phases mapped onto the ternary diagram. We have eliminated here the unfavorable compositions in the corresponding phases in accordance with Figure 2.

As can be seen from Figure 3a,b, the formation energy is negative for the majority of compounds. The maps of E_{form} for favorable austenitic and martensite structures are similar. For both cases, the most stable compounds are close to Ni$_{50}$Mn$_{25}$Ga$_{25}$ in concentration ranges of $45 < x < 65$ at.%Ni, $6 < y < 30$ at.%Mn, and $18 < z < 45$ at.%Ga.

In general, the supercell calculations with structural relaxations reproduce the corresponding static relaxation results for Ni–Mn–Ga in the austenitic phase [37–39]. However, comparative analysis of the two approaches shows that the Ga- and Mn-rich compounds located at the left and top corners of the ternary map turn out to be unstable ($E_{form} > 0$) after the full geometric optimization as compared to the static relaxation procedure, which predict the Ga-rich compositions to be unstable only in the austenitic phase [37,39]. In the martensitic phase, among 92 compositions, only two (Ni$_{6.25}$Mn$_{12.5}$Ga$_{81.25}$ and Ni$_{6.25}$Mn$_{18.75}$Ga$_{75}$) orthorhombic structures have positive formation energies and decompose into the elements involved. We have excluded these compositions from further consideration.

The criterion of stability in terms of formation energy is not rigorous since some multicomponent compounds can decompose into stable binary and ternary prototypes together with elemental components. Therefore, it is necessary to consider the thermodynamic phase stability of Ni$_x$Mn$_y$Ga$_z$ in terms of the energy convex hull analysis. The main idea is to compare E_{form} of the ternary compound under study with a possible combination of stable phases (pivot points). The three-dimensional convex hull and its cross-sections into the formation energy-composition space of Ni–Mn–Ga with austenite and martensite phases are shown in Figure S5 of the SM. The majority of investigated compounds have negative formation energies but lie above or closely above the convex hull. This finding indicates that these compositions are in a thermodynamically metastable austenitic or martensitic phase and can decompose.

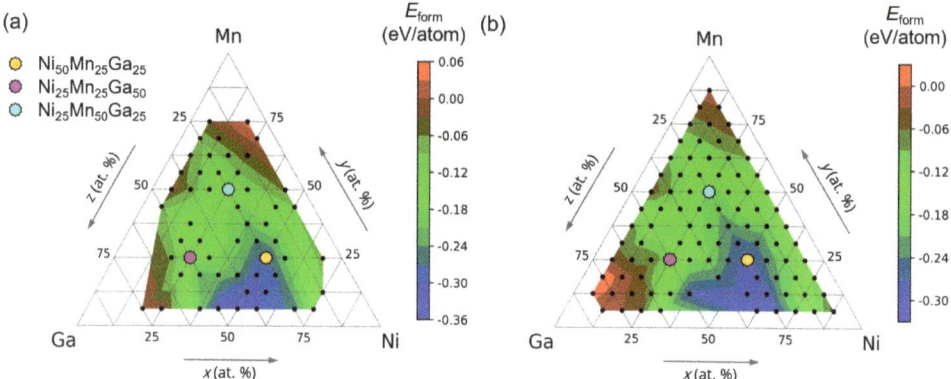

Figure 3. (Color online) Formation energy mapped into the ternary diagram of $Ni_xMn_yGa_z$ compounds with favorable (**a**) cubic and (**b**) tetragonal and orthorhombic crystal structures.

Figures 4a,b display contour maps of the decomposition reactions indicating the stable, metastable, and unstable $Ni_xMn_yGa_z$ compositions in the austenite and martensite. One can see that for both phases, the Ga-rich compounds concentrated near the left corner of the ternary diagram are unstable and reveal a tendency for decomposing into stable sub-compounds as suggested by the convex hull. It happens because the Ga-rich compounds lie sufficiently above the convex hull. Compounds from the middle part of the diagram are found to be metastable with a fraction of decomposition reactions at about 50%. Finally, compounds of $z < 30$ at.%Ga located at the right side of the ternary diagram are potentially stable or nearly stable (i.e., lying on or right above the convex hull).

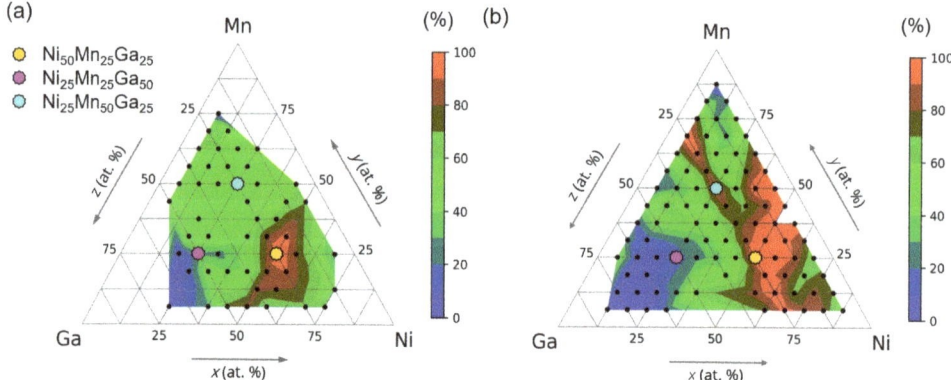

Figure 4. (Color online) The ternary maps of the stable reactions for $Ni_xMn_yGa_z$ in the (**a**) austenitic and (**b**) martensitic phases against the decomposition into a mixture of stable compounds. Here, 0% and 100% correspond to the unstable and stable $Ni_xMn_yGa_z$ compositions.

In particular, the fraction of compounds with the stable tetragonal structure is sufficiently greater than those with the stable cubic structure. We note that the Ni–Mn–Ga compounds predicted to be stable in the austenite phase are concentrated near the stoichiometric Ni$_2$MnGa. This finding reproduces the experimental ternary diagram well, revealing the most studied Ni–Mn–Ga compounds (see Figure S6 in the SM).

Figure 5a illustrates the contour map of the optimized lattice constant a_0 of the cubic austenite Ni$_x$Mn$_y$Ga$_z$. It can be seen that equilibrium lattice parameters depend mostly on the Ga content. Compounds with Ga excess $z > 55$ at.% exhibit the large $a_0 > 6$ Å due to the larger atomic radius of Ga as compared to those of Ni and Mn. A gradual decrease in Ga doping leads to a reduction in a_0. For compositions with $x < 55$ at.%Ni and $25 < z < 55$ at.%Ga, a_0 takes values in the range between 5.85 and 5.95 Å, whereas the further reduction of Ga at the same Ni content decreases the lattice parameter to $a_0 \approx 5.78$ Å. A sufficiently smaller lattice constant (≈ 5.68 Å) is observed for Ni-rich compositions with $x > 70$ at.%Ni.

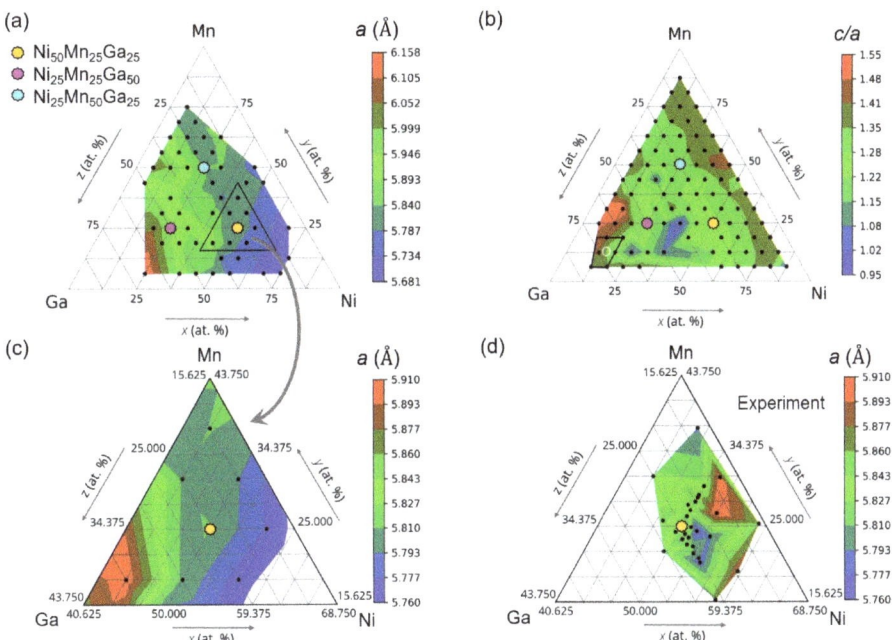

Figure 5. (Color online) Calculated (**a**) equilibrium lattice parameter a_0 and (**b**) tetragonal ratio c/a mapped into the ternary diagram of Ni$_x$Mn$_y$Ga$_z$ compounds in austenite and martensite. (**c**) Theoretical and (**d**) experimental large-scale a_0 maps. The list of experimental compositions is tabulated in Table S17, see the SM. Label O in (**b**) denotes the region with a favorable orthorhombic structure. For the orthorhombic structure, the b/a ratio is presented in Table S8, see the SM.

In Figure 5b, we display the contour map of the tetragonal ratio c/a of martensitic Ni$_x$Mn$_y$Ga$_z$. For Mn- or Ni-rich compounds with the Ga content of $z < 20$ at.%, the tetragonal structure is characterized by the ratios $1.3 < c/a < 1.4$. An increase in both Ga and Ni or Ga and Mn concentrations simultaneously leads to a reduction in the c/a range to $1.2 < c/a < 1.3$. The quantity of these compounds is the largest and they are represented within the middle area of diagram with the exception of compositions with $25 < x < 57$ at.%Ni, $y < 35$ at.%Mn, and $31 < z < 57$ at.%Ga (see the blue map in Figure 5b), which are stable in the cubic phase ($c/a = 1$) only. The largest tetragonal ratios varied in the range of $1.4 < c/a < 1.5$ are found for tetragonally distorted Ga-rich compounds ($x < 18$ at.%, $18 < y < 30$ at.%, and $55 < z < 70$ at.%). We would like to mention that for Ga-rich compounds ($x < 18$ at.%, $y < 18$ at.%, and $z > 64$ at.%), the orthorhombic structure with $0.9 < b/a < 1$ and $1.3 < c/a < 1.4$ is favorable. The calculated lattice constants for both austenitic and martensitic structures are summarized in the SM (see Tables S5–S10).

Our data allow for a systematic comparison of the calculated lattice parameters with available experimental data. The calculated equilibrium lattice constant a_0 of the stoichiometric compositions are ≈5.81 Å for Ni_2MnGa, 5.84 Å for Mn_2NiGa, and 5.95 Å for Ga_2MnNi. These values are close to experimental ones: 5.82 Å for Ni_2MnGa [1], 5.9 Å for Mn_2NiGa [23], and 5.84 Å for Ga_2MnNi [29]. Despite the slight difference in theoretical and experimental values for Mn_2NiGa and Ga_2MnNi, it should be pointed out that our results are found to be in excellent agreement with other ab initio studies. For instance, Barman et al. [29] reported $a_0 = 5.96$ Å for Ga_2MnNi and Kundu et al. [28] obtained 5.84 Å for Mn_2NiGa.

As for the off-stoichiometric compositions, there is a limited list of compounds studied experimentally, and most of them are concentrated near Ni_2MnGa (see Figure S6 in the SM). Figure 5c,d show magnifications of the theoretical and experimental data in the vicinity of Ni_2MnGa maps of a_0. The lattice constants of the most experimental compounds take values between 5.81 and 5.86 Å, except the $Ni_{51.2}Mn_{31.1}Ga_{17.7}$ and $Ni_{52}Mn_{24.4}Ga_{23.6}$ with the largest $a_0 = 5.91$ Å and lowest $a_0 = 5.76$ Å [45], respectively. The discrepancy between experimental and calculated values is ≈1%. Apart from the accuracy of ab initio simulations, the difference could arise from the fact that the experimental values of a_0 were obtained at finite temperatures (i.e., at temperatures above T_m). For instance, T_m for $Ni_{51.2}Mn_{31.1}Ga_{17.7}$ with the largest a_0 is about 446 K [45]. In contrast, ab initio methods assume zero temperatures and lattice constants have to be slightly less. As Figure 5d suggests, the lattice constant reduces and reaches the value of about 5.76 Å with increasing (decreasing) Ni (Mn) content at the fixed ≈25 at.%Ga. On the other hand, the successive replacement of Ga by Mn and vice versa at the fixed 50 at.%Ni does not change the lattice constant sufficiently, which is about 5.84 Å. The similar trends for $Ni_{50\pm x}Mn_{25\pm y}Ga_{25}$ and $Ni_{50}Mn_{25\pm y}Ga_{25\pm z}$ can be seen in the theoretical a_0 map, as shown in Figure 5c.

Total magnetic moments (μ_{tot}) for stable compositions in the austenite and martensite are shown in contour maps of Figures 6a,b, correspondingly, where we also marked areas with favorable magnetic configurations. One can see a similarity in the magnetic moments of the austenitic and martensitic phases. Namely, Ga- and Ni-rich compounds (area I) are ordered ferromagnetically. Ga-rich compositions have a small magnetic moment less than 1.5 μ_B/f.u.; which is correlated with larger contents of non-magnetic Ga diluting the ferromagnetically ordered phase. Compositions of area II are predominantly ordered ferrimagnetically with FIM-1 type. At the same time, there are regions with FIM-2, FIM-1(2), and FM spin configurations on the right and left sides of the area II. We note that inverted spins are assigned to Mn-excess atoms located at Ni sites for FIM-1 and at Ga sites for FIM-2. This effect takes place for all Mn excess atoms in the case of the FIM-1(2) state. The largest magnetic moments of 5.5 μ_B/f.u. and 4.52 μ_B/f.u. are obtained for FM austenite $Ni_{62.5}Mn_{31.25}Ga_{6.25}$ and FM martensite $Ni_{68.75}Mn_{25}Ga_{6.25}$. In addition, the largest difference between μ_{tot} in austenite and martensite is about 2.5 μ_B/f.u. for composition $50 < x < 75$ at.%Ni and $z < 25$ at.%Ga.

Let us consider the Mn-rich compounds mapped in the area III. Since Mn atoms can occupy four fcc sublattices constituting the inverse Heusler structure depending on nominal composition, several FIM alignments can be realized. Among the seven considered spin configurations, FIM-3 and FIM-7 are predicted to be favorable. For FIM-3, Mn atoms at the 4a site have reversed spin alignment, while Mn atoms at both 4a and 4b have opposite spins in the case of FIM-7. The total magnetic moment is smaller than 1.5 μ_B/f.u particularly in the tetragonal martensitic phase because of the complex magnetic order. The largest difference in total magnetic moments between the austenite and martensite is ≈2.5 μ_B/f.u. for compounds with $10 < x < 35$ at.% and $z < 25$ at.%. The partial magnetic moments for all compositions are listed in Tables S11–S15 of the SM.

In contrast to Reference [39], the relaxation of the atomic positions allows to reveal distinguishing features of the magnetic phase diagram, which are: (i) the FIM-1 phase with a larger magnetic moment (We would like to note that there is the misprint in Figure 3a of Ref. [39]. The unit for total magnetic moment plotted in the area II of diagram is μ_B/atom instead μ_B/f.u. As consequence, there is a difference between magnetic moments calculated within the full geometric and static relaxation procedures.) exists for compounds with $y < 50$ at.%Mn and (ii) the FIM-3 phase occurs for compounds with $y \geq 50$ at.%Mn.

In Figure 6c,d, we illustrate the zoomed-in contour maps of μ_{tot} for a set of theoretical and experimental compounds in the martensitic phase for compositions near $Ni_{50}Mn_{25}Ga_{25}$. Close inspection of the data shows that the theoretical μ_{tot} correlates with experimental ones. In both cases, the largest $\mu_{tot} \approx 4$ μ_B/f.u. is found for $Ni_{50}Mn_{25}Ga_{25}$, whereas deviations from the stoichiometry such as $Ni_{50\pm x}Mn_{25\mp x}Ga_{25}$, $Ni_{50}Mn_{25\pm x}Ga_{25\mp x}$, and $Ni_{50\pm x}Mn_{25}Ga_{25\mp x}$ result in the magnetic moment reduction to 2 μ_B/f.u. For experimental samples $46 < x < 57$ at.%Ni, $y < 19$ at.%Mn, and $28 < z < 35$ at.%Ga, the magnetic moment is less than 1 μ_B/f.u. Importantly, these calculations do not predict any compounds with a favorable tetragonal structure in the present concentration range. Such distinction can be caused by the presence of modulated martensitic structure in the experimental compounds, while modulations were not taken into account in calculations due to an increase in computational problems.

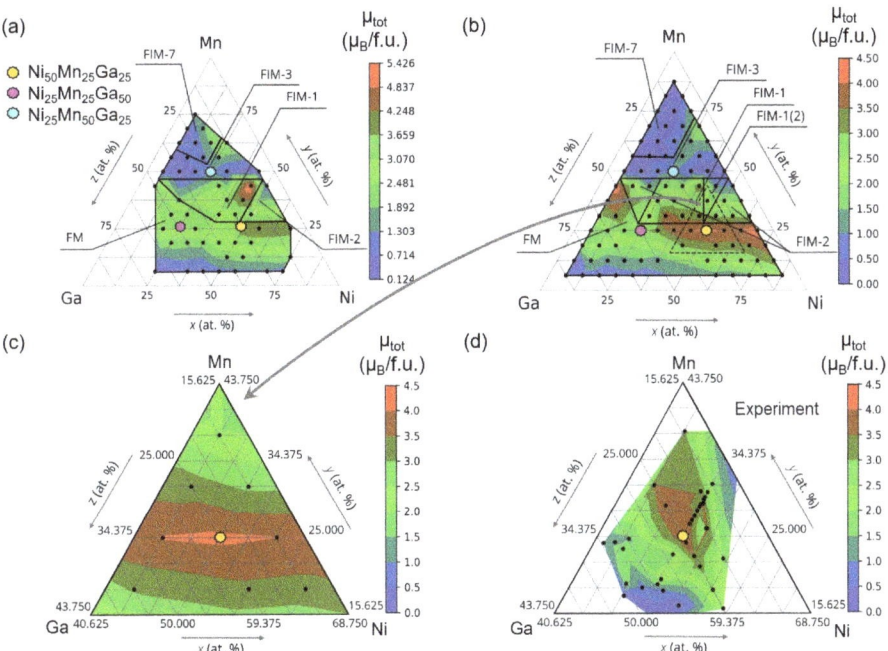

Figure 6. (Color online) Calculated total magnetic moments together with preferable spin alignment of $Ni_xMn_yGa_z$ in the (**a**) austenitic and (**b**) martensitic phase. (**c**) Theoretical and (**d**) experimental large-scale maps of the total magnetic moment in the martensitic phase. The experimental compositions are listed in Table S18, see the Supplementary Materials.

Having the total energies of the austenitic and martensitic phases, we can estimate the martensitic transformation temperature T_m using the relation $\Delta E \approx k_B T_m$, where ΔE is the energy difference between the cubic and tetragonal structures and k_B is the Boltzmann constant. Obtained T_m for $Ni_xMn_yGa_z$ are presented in Figure 7. One can see that the martensitic transformation is predicted for $\approx 40\%$ of compounds under study, most of those are compositions with $z < 30$ at.%Ga. The lowest T_m is obtained for compositions with

$25 < z < 60$ at.%Ni, $15 < y < 30$ at.%Mn, and $25 < z < 55$ at.%Ga, while the largest ones are found for Mn-rich compounds with a small Ga content. In particular, we predict $T_m \approx 1160$ K for $Ni_{43.75}Mn_{50}Ga_{6.25}$. This compound is close in composition to a binary NiMn, which demonstrates the B2-L1$_0$ martensitic transformation at about 1000 K [46,47].

Figure 7b,c compare the theoretical and experimental zoomed-in contour maps of martensitic for compositions concentrated near $Ni_{50}Mn_{25}Ga_{25}$. One can see a noticeable correlation between our results and experimental results. For the $Ni_{50}Mn_{25\pm x}Ga_{25\mp x}$ compound, the subsequent replacement of Ga by Mn atoms leads to an increase in T_m. The theoretically obtained change in T_m of about 150 K between $Ni_{50}Mn_{25}Ga_{25}$ and $Ni_{50}Mn_{31.25}Ga_{18.75}$ is in agreement with an experimental one between $Ni_{50}Mn_{25}Ga_{25}$ and $Ni_{50}Mn_{28.9}Ga_{21.1}$ [22]. A similar trend can be observed for T_m of $Ni_{50+x}Mn_{25}Ga_{25-x}$ and $Ni_{50+x}Mn_{25-x}Ga_{25}$ but with a smaller change of T_m with x. On the other hand, an increase in Ga content leads to T_m reduction as it follows from theoretical and experimental ternary maps. Since substitution of Mn or Ni for Ga provides a change in the e/a ratio, it results in an increasing T_m. Figure 7a,b show that the majority of considered compounds demonstrate this tendency and reproduce the experimental $T_m(e/a)$ behavior [17,20,22] in accordance with the Hume–Rothery mechanism. A correlation between Figures 5b and 7 allows us to conclude that a high martensitic transition temperature is observed for compositions with a large tetragonality (c/a ratio) of the martensite. A detailed inspection of Figure 7b,c suggests that the theoretical values of T_m are calculated to be less than the experimental ones. This disagreement can be minimized by an extension of zero-temperature ab initio calculations to finite temperatures in terms of the balance of free energies in the different magnetic, premartensitic, and martensitic phases as proposed by Dutta et al. [35].

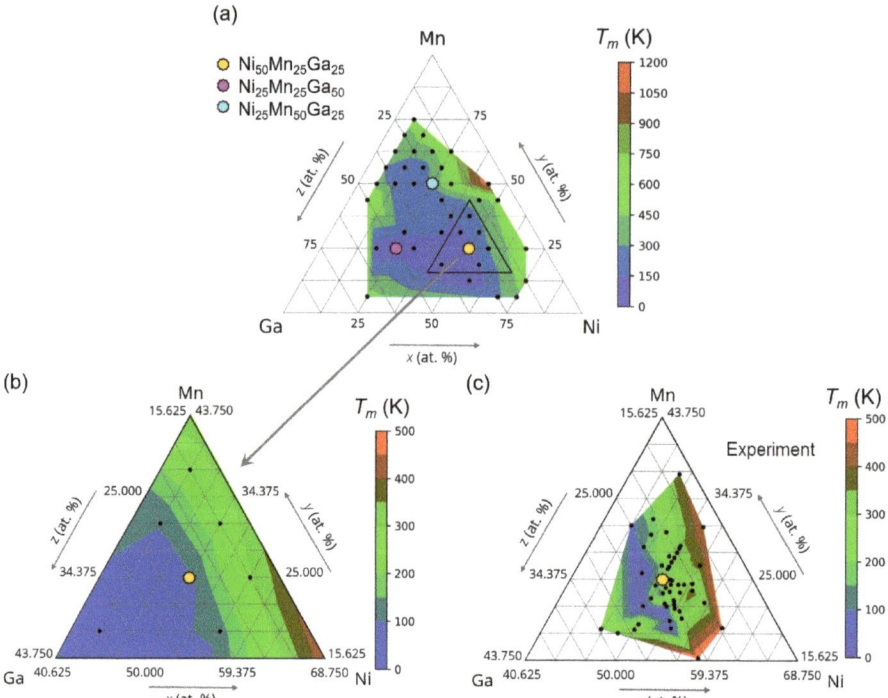

Figure 7. (Color online) (**a**,**b**) Theoretical and (**c**) experimental martensitic transition temperature of $Ni_xMn_yGa_z$. In the case of (**b**,**c**), the large-scale contour maps in the vicinity of $Ni_{50}Mn_{25}Ga_{25}$ are illustrated. The list of experimental compositions is presented in Table S19, see the SM.

4. Conclusions

In summary, we carried out a systematic investigation of Ni–Mn–Ga austenite and martensite with respect to their magnetic and structural properties. A mesh of 105 Ni–Mn–Ga compositions was studied using the first-principles supercell calculations. The ground state atomic and magnetic configurations for cubic austenitic and tetragonal martensitic phases were simulated and considered based on their chemical and structural stability. It is shown that 51 compositions located mainly on the left ($z > 45$ at.%Ga) and right ($y > 45$ at.%Mn) sides of triangle diagrams are unstable in the cubic austenite. In the case of the martensitic phase, the majority of compositions prefer the tetragonal structure with the c/a ratio varied from 1.15 to 1.5. The exception is the compounds $25 < x < 57$ at.%Ni, $y < 35$ at.%Mn, and $31 < z < 57$ at.%Ga, which prefer the cubic structure. Ga-rich compounds have an orthorhombic structure. The phase stability analysis conducted in terms of the convex hull showed that (i) Ga-rich compositions with $z > 60$ at.%Ga decompose with utmost probability into stable pure components and binary phases that lie on the convex hull; (ii) compounds with $35 < z < 60$ at.%Ga tend to decompose with a probability of about 50%; (iii) alloys with $z < 35$ at.%Ga are stable against decomposition, and their phase stability rises with decreasing Ga content.

The magnetic phase diagrams are found to be similar for austenite and martensite, revealing the regions of FM and FIM spin alignment. The ferromagnetically ordered austenitic and martensitic phases are available predominantly in compounds with Mn content $y < 30$ at.%. In contrast, the complex FIM order with various types of Mn spin alignment takes place for the compositions with $y > 30$ at.%Mn. Alloys with $10 < x < 35$ at.%Ni and $z < 25$ at.%Ga reveal the largest magnetization difference between the austenitic and martensitic phases, which turned out to be ≈ 2.5 μ_B/f.u.

Overall, this work demonstrates the predictive power of first-principles calculations, which can be used for the heuristic search of functional magnetic materials and their characterization and for the prediction of compounds with novel functionalities. Despite limitations in ab initio methods, results obtained in this work demonstrate our improved understanding of what it takes to predict stable martensitic off-stoichiometric magnetic compounds. With the increasing expertise in this field, one could undertake a full-feature study with larger supercells, allowing for various types of structural degrees of freedom, like the modulations in the case of Ni–Mn–Ga.

Supplementary Materials: The following are available online at https://www.mdpi.com/article/10.3390/met11060973/s1, Figure S1: 16-atom cubic supercells for stoichiometric Ni–Mn–Ga with L2$_1$ crystal structure. (a) Ni$_8$Mn$_4$Ga$_4$ ($Fm\bar{3}m$, #225); (b) Ga$_8$Mn$_4$Ni$_4$ ($Fm\bar{3}m$, #225′); (c) Mn$_8$Ni$_4$Ga$_4$ ($Fm\bar{3}m$, #225″); (d) Mn$_8$Ni$_4$Ga$_4$ ($F\bar{4}3m$, #216). Here, Ni, Mn (Mn$_1$ for #216 structure), Ga, and Mn$_2$ (for #216 structure) atoms are denoted by green, red, light blue, and orange colors, respectively; Figure S2: 16-atom cubic supercells for off-stoichiometric Ni–Mn–Ga with #225^{mix} structure: (a) Ni$_7$Mn$_4$Ga$_5$ and (b) Ni$_5$Mn$_4$Ga$_7$. Here Ni, Mn, Ga, and Ni-excess atoms are denoted by green, red, light blue, and orange colors, respectively; Figure S3: Structures in a cubic space group #225 for Ni–Mn–Ga with (a) FM, (b) FIM-1, and (c) FIM-2 spin configurations. Here, Ni, Mn, Ga, and Mn-excess atoms are denoted by green, red, light blue, and violet colors, respectively; Figure S4: Structures in a cubic space group #216 for Ni–Mn–Ga with (a) FIM-3, (b) FIM-7 spin configurations. Here Ni, Mn$_1$, Mn$_2$ Ga, and Mn-excess atoms are denoted by green, red, orange, light blue, and violet colors, respectively; Figure S5: The three-dimensional convex hull in the formation energy of Ni–Mn–Ga compositions with the (a) austenite and (b) martensite phases. (c), (d) The cross-sections of convex hull plotted at the fixed Mn content (y = 25, 50, and 75 at.%). Here lines denote the profiles of the convex hull. The degree of stable reactions among the possible ones for a particular compound against the decomposition into a mixture of stable phases is also indicated above symbols; Figure S6: Compositional ternary phase diagram of Ni–Mn–Ga compounds from (Sokolovskaya Yu.A.; Sokolovskiy V.V.; Zagrebin M.A.; Buchelnikov V.D.; Zayak A.T. Ab Initio Study of the Composite Phase Diagram of Ni–Mn–Ga Shape Memory Alloys. *J. Exp. Theor. Phys.* **2017**, *152*, 125–132.). The points map the most studied compositions (about 900) reported in the literature; Table S1: Positions of the Ni, Mn and Ga atoms considered in the regular (space group #225) and inverse (space group #216) Heusler structures;

Table S2: The considered orientation of spin magnetic moment of Mn atoms located at different Wyckoff positions in the regular (#225) and inverse (#216) Heusler structures. Here the reversed spin orientation is marked in a red color to enhance visibility. Notice that the Ni spin magnetic moment is aligned along the direction ⇑; Table S3: Space group, lattice parameter a (in Å), c/a ratio, and total energy E_{tot} (in eV/atom) per atom for the pivot points of the ternary Ni–Mn–Ga convex hull; Table S4: Possible decomposition reactions and the decomposition energy E_{dec} (in eV/atom) between the investigated alloy and reaction products for $Ni_9Mn_4Ga_3$. The positive value indicates the phase instability against the decomposition process and vice versa; Table S5: Optimized lattice constants a, b, c (in Å) and their ratios, the total energy E_{tot} and the formation energy E_{form} (in eV/atom) as well as the preferable structure for compounds in the austenite phase from the area I; Table S6: Optimized lattice constants a, b, c (in Å) and their ratios, the total energy E_{tot} and the formation energy E_{form} (in eV/atom) as well as the preferable structure for compounds in the austenite phase from the area II; Table S7: Optimized lattice constants a, b, c (in Å) and their ratios, the total energy E_{tot} and the formation energy E_{form} (in eV/atom) as well as the preferable structure for compounds in the austenite phase from the area III; Table S8: Optimized lattice constants a, b, c (in Å) and their ratios, the total energy E_{tot} and the formation energy E_{form} (in eV/atom) for compounds in the martensite phase from the area I; Table S9: Optimized lattice constants a, b, c (in Å) and their ratios, the total energy E_{tot} and the formation energy E_{form} (in eV/atom) for compounds in the martensite phase from the area II; Table S10: Optimized lattice constants a, b, c (in Å) and their ratios, the total energy E_{tot} and the formation energy E_{form} (in eV/atom) for compounds in the martensite phase from the area III; Table S11: Element resolved magnetic moments (in μ_B), total magnetic moments (in μ_B/f.u.) and favorable magnetic reference state for compounds in the austenite and martensite phase from the area I; Table S12: Element resolved magnetic moments (in μ_B), total magnetic moments (in μ_B/f.u.) and favorable magnetic reference state for compounds in the austenite phase from the area II. Here $\mu_{Mn^{(Ga)}}$ and $\mu_{Mn^{(Ni)}}$ are the magnetic moments of Mn atoms placed at the Ga- and Ni sublattice, respectively; Table S13: Element resolved magnetic moments (in μ_B), total magnetic moments (in μ_B/f.u.) and favorable magnetic reference state for compounds in the martensite phase from the area II. Here $\mu_{Mn^{(Ga)}}$ and $\mu_{Mn^{(Ni)}}$ are the magnetic moments of Mn atoms placed at the Ga- and Ni sublattice, respectively; Table S14: Element resolved magnetic moments (in μ_B), total magnetic moments (in μ_B/f.u.) and favorable magnetic reference state for compounds in the austenite phase from the area III. Here $\mu_{Mn^{(4a)}}$ and $\mu_{Mn^{(4b)}}$ are the magnetic moments of Mn atoms, which occupy $4a$ and $4b$ Wyckoff sites while $\mu_{Mn^{(Ga)}}$ and $\mu_{Mn^{(Ni)}}$ are the Mn magnetic moments at the Ga- and Ni sites, respectively; Table S15: Element resolved magnetic moments (in μ_B), total magnetic moments (in μ_B/f.u.) and favorable magnetic reference state for compounds in the martensite phase from the area III. Here $\mu_{Mn^{(4a)}}$ and $\mu_{Mn^{(4b)}}$ are the magnetic moments of Mn atoms, which occupy $4a$ and $4b$ Wyckoff sites while $\mu_{Mn^{(Ga)}}$ and $\mu_{Mn^{(Ni)}}$ are the Mn magnetic moments at the Ga- and Ni sites, respectively; Table S16: Martensitic transition temperature T_m (in K) for compositions from the areas I, II, and III; Table S17: Experimental value of the lattice constants a (in Å) for austenitic phase; Table S18: Experimental value of the total magnetic moments (in μ_B/f.u.) for compounds in the martensite phase; Table S19: Experimental value of the martensitic transition temperature T_m (in K).

Author Contributions: Y.S.: ab initio calculations; O.M.: ab initio calculations, writing; D.B.: calculations of convex hull energy and decomposition reactions, visualization; V.S.: conceptualization, methodology, visualization—review and editing; M.Z.: methodology, ab initio calculations, writing, visualization; V.B.: conceptualization, methodology, writing—review and editing, supervision; A.T.Z.: conceptualization, methodology. All authors have read and agreed to the published version of the manuscript.

Funding: The work was carried out with financial support from the Ministry of Science and Higher Education of the Russian Federation in the framework of Increase Competitiveness Program of NUST "MISiS" (No. K2-2020-045), implemented by a governmental decree dated 16 March 2013, No. 211.

Institutional Review Board Statement: Not applicable.

Informed Consent Statement: Not applicable.

Data Availability Statement: Not applicable.

Conflicts of Interest: The authors declare no conflict of interest.

References

1. Webster, P.J.; Ziebeck, K.R.A.; Town, S.L.; Peak, M.S. Magnetic order and phase transformation in Ni$_2$MnGa. *Philos. Mag. B* **1984**, *49*, 295–310. [CrossRef]
2. Ullakko, K.; Huang, J.K.; Kantner, C.; O'Handley, R.C. Large magnetic-field-induced strains in Ni$_2$MnGa single crystals. *Appl. Phys. Lett.* **1996**, *69*, 1966–1968. [CrossRef]
3. Chernenko, V.A.; Seguí, C.; Cesari, E.; Pons, J.; Kokorin, V.V. Sequence of martensitic transformations in Ni-Mn-Ga alloys. *Phys. Rev. B* **1998**, *57*, 2659–2662. [CrossRef]
4. Wedel, B.; Suzuki, M.; Murakami, Y.; Wedel, C.; Suzuki, T.; Shindo, D.; Itagaki, K. Low temperature crystal structure of Ni-Mn-Ga alloys. *J. Alloys Compd.* **1999**, *290*, 137–142. [CrossRef]
5. Glavatska, N.; Mogilniy, G.; Glavatsky, I.; Danilkin, S.; Hohlwein, D.; Beskrovnij, A.; Lindroos, V.K. Temperature dependence of martensite structure and its effect on magnetic-field-induced strain in Ni$_2$MnGa magnetic shape memory alloys. *J. Phys. IV France* **2003**, *112*, 963–967. [CrossRef]
6. Gavriljuk, V.G.; Söderberg, O.; Bliznuk, V.V.; Glavatska, N.I.; Lindroos, V.K. Martensitic transformations and mobility of twin boundaries in Ni$_2$MnGa alloys studied by using internal friction. *Scripta Mater.* **2003**, *49*, 803–809. [CrossRef]
7. Vasil'ev, A.N.; Buchel'nikov, V.D.; Takagi, T.; Khovailo, V.V.; Estrin, E.I. Shape memory ferromagnets. *Phys. Usp.* **2003**, *46*, 559–588. [CrossRef]
8. Dubenko, I.; Samanta, T.; Pathak, A.K.; Kazakov, A.; Prudnikov, V.; Stadler, S.; Granovsky, A.; Zhukov, A.; Ali, N. Magnetocaloric effect and multifunctional properties of Ni-Mn-based Heusler alloys. *J. Magn. Magn. Mater.* **2012**, *324*, 3530–3534. [CrossRef]
9. Dubenko, I.; Quetz, A.; Pandey, S.; Aryal, A.; Eubank, M.; Rodionov, I.; Prudnikov, V.; Granovsky, A.; Lähderanta, E.; Samanta, T.; et al. Multifunctional properties related to magnetostructural transitions in ternary and quaternary Heusler alloys. *J. Magn. Magn. Mater.* **2015**, *383*, 186–189. [CrossRef]
10. Sozinov, A.; Likhachev, A.A.; Lanska, N.; Ullakko, K. Giant magnetic-field-induced strain in NiMnGa seven-layered martensitic phase. *Appl. Phys. Lett.* **2002**, *80*, 1746–1748. [CrossRef]
11. Buchelnikov, V.D.; Vasiliev, A.N.; Koledov, V.; Taskaev, S.V.; Khovaylo, V.V.; Shavrov, V.G. Magnetic shape-memory alloys: Phase transitions and functional properties. *Phys. Usp.* **2006**, *49*, 871–877. [CrossRef]
12. Planes, A.; Manosa, L.; Acet, M. Magnetocaloric effect and its relation to shape-memory properties in ferromagnetic Heusler alloys. *J. Phys. Condens. Matter* **2009**, *21*, 233201. [CrossRef]
13. Kamantsev, A.P.; Koledov, V.V.; Mashirov, A.; Dilmieva, E.T.; Shavrov, V.G.; Cwik, J.; Los, A.S.; Nizhankovskii, V.I.; Rogacki, K.; Tereshina, I.S.; et al. Magnetocaloric and thermomagnetic properties of Ni$_{2.18}$Mn$_{0.82}$Ga Heusler alloy in high magnetic fields up to 140 kOe. *J. Appl. Phys.* **2015**, *117*, 163903. [CrossRef]
14. Pons, J.; Chernenko, V.A.; Santamarta, R.; Cesari, E. Crystal structure of martensitic phases in Ni-Mn-Ga shape memory alloys. *Acta Mater.* **2000**, *48*, 3027–3038. [CrossRef]
15. Bozhko, A.D.; Vasil'ev, A.N.; Khovailo, V.V.; Buchel'nikov, V.D.; Dikshtein, I.E.; Seletskii, S.M.; Shavrov, V.G. Phase transitions in the ferromagnetic alloys Ni$_{2+x}$Mn$_{1-x}$Ga. *JETP Lett.* **1998**, *67*, 227–232. [CrossRef]
16. Vasil'ev, A.N.; Bozhko, A.D.; Khovailo, V.V.; Dikshtein, I.E.; Shavrov, V.G.; Buchelnikov, V.D.; Matsumoto, M.; Suzuki, S.; Takagi, T.; Tani, J. Structural and magnetic phase transitions in shape-memory alloys Ni$_{2+x}$Mn$_{1-x}$Ga. *Phys. Rev. B* **1999**, *59*, 1113–1120. [CrossRef]
17. Jin, X.; Marioni, M.; Bono, D.; Allen, S.M.; O'Handley, R.C.; Hsu, T.Y. Empirical mapping of Ni-Mn-Ga properties with composition and valence electron concentration. *J. Appl. Phys.* **2002**, *91*, 8222–8224. [CrossRef]
18. Khovaylo, V.V.; Buchelnikov, V.D.; Kainuma, R.; Koledov, V.V.; Otsuka, M.; Shavrov, V.G.; Takagi, T.; Taskaev, S.V.; Vasiliev, A.N. Phase transitions in Ni$_{2+x}$Mn$_{1-x}$Ga with a high Ni excess. *Phys. Rev. B* **2005**, *72*, 224408. [CrossRef]
19. Entel, P.; Buchelnikov, V.D.; Khovailo, V.V.; Zayak, A.T.; Adeagbo, W.A.; Gruner, M.E.; Herper, H.C.; Wassermann, E.F. Modelling the phase diagram of magnetic shape memory Heusler alloys. *J. Phys. D Appl. Phys.* **2006**, *39*, 865–889. [CrossRef]
20. Wang, Y.F.; Wang, J.M.; Jiang, C.B.; Xu, H.B. Composition triangle diagrams of Ni-Mn-Ga magnetic shape memory alloys. *Acta Metall. Sin. (Engl. Lett.)* **2006**, *19*, 171–175. [CrossRef]
21. Buchelnikov, V.D.; Taskaev, S.V.; Zagrebin, M.A.; Ermakov, D.I.; Koledov, V.V.; Shavrov, V.G.; Takagi, T. The phase diagrams of Ni-Mn-Ga alloys in the magnetic field. *J. Magn. Magn. Mater.* **2007**, *313*, 312–316. [CrossRef]
22. Çakır, A.; Righi, L.; Albertini, F.; Acet, M.; Farle, M.; Aktúrk, S. Extended investigation of intermartensitic transitions in Ni-Mn-Ga magnetic shape memory alloys: A detailed phase diagram determination. *J. Appl. Phys.* **2013**, *114*, 183912. [CrossRef]
23. Liu, G.D.; Chen, J.L.; Liu, Z.H.; Dai, X.F.; Wu, G.H.; Zhang, B.; Zhang, X.X. Martensitic transformation and shape memory effect in a ferromagnetic shape memory alloy: Mn$_2$NiGa. *Appl. Phys. Lett.* **2005**, *87*, 262504. [CrossRef]
24. Liu, G.D.; Dai, X.F.; Yu, S.Y.; Zhu, Z.Y.; Chen, J.L.; Wu, G.H.; Zhu, H.; Xiao, J.Q. Physical and electronic structure and magnetism of Mn$_2$NiGa: Experiment and density-functional theory calculations. *Phys. Rev. B* **2006**, *74*, 054435. [CrossRef]
25. Singh, S.; Maniraj, M.; D'Souza, S.W.; Ranjan, R.; Barman, S.R. Structural transformations in Mn$_2$NiGa due to residual stress. *Appl. Phys. Lett.* **2010**, *96*, 081904. [CrossRef]
26. Barman, S.R.; Banik, S.; Shukla, A.K.; Kamal, C.; Chakrabarti, A. Martensitic transition, ferrimagnetism and Fermi surface nesting in Mn$_2$NiGa. *Europhys. Lett.* **2007**, *80*, 57002. [CrossRef]
27. Chakrabarti, A.; Siewert, M.; Roy, T.; Mondal, K.; Banerjee, A.; Gruner, M.E.; Entel, P. Ab initio studies of effect of copper substitution on the electronic and magnetic properties of Ni$_2$MnGa and Mn$_2$NiGa. *Phys. Rev. B* **2013**, *88*, 174116. [CrossRef]

28. Kundu, A.; Gruner, M.E.; Siewert, M.; Hucht, A.; Entel, P.; Ghosh, S. Interplay of phase sequence and electronic structure in the modulated martensites of Mn$_2$NiGa from first-principles calculations. *Phys. Rev. B* **2017**, *96*, 064107-13. [CrossRef]
29. Barman, S.R.; Chakrabarti, A.; Singh, S.; Banik, S.; Bhardwaj, S.; Paulose, P.L.; Chalke, B.A.; Panda, A.K.; Mitra, A.; Awasthi, A.M. Theoretical prediction and experimental study of a ferromagnetic shape memory alloy: Ga$_2$MnNi. *Phys. Rev. B* **2008**, *78*, 134406. [CrossRef]
30. Uijttewaal, M.A.; Hickel, T.; Neugebauer, J.; Gruner, M.E.; Entel, P. Understanding the Phase Transitions of the Ni$_2$MnGa Magnetic Shape Memory System from First Principles. *Phys. Rev. Lett.* **2009**, *102*, 035702. [CrossRef]
31. Buchelnikov, V.D.; Sokolovskiy, V.V.; Herper, H.C.; Ebert, H.; Gruner, M.E.; Taskaev, S.V.; Khovaylo, V.V.; Hucht, A.; Dannenberg, A.; Ogura, M.; et al. First-principles and Monte Carlo study of magnetostructural transition and magnetocaloric properties of Ni$_{2+x}$Mn$_{1-x}$Ga. *Phys. Rev. B* **2010**, *81*, 094411. [CrossRef]
32. Li, C.-M.; Luo, H.-B.; Hu, Q.-M.; Yang, R.; Johansson, B.; Vitos, L. Temperature dependence of elastic properties of Ni$_{2+x}$Mn$_{1-x}$Ga and Ni$_2$Mn(Ga$_{1-x}$Al$_x$) from first principles. *Phys. Rev. B* **2011**, *84*, 174117. [CrossRef]
33. Hickel, T.; Uijttewaal, M.; Al-Zubi, A.; Dutta, B.; Grabowski, B.; Neugebauer, J. Ab Initio-Based Prediction of Phase Diagrams: Application to Magnetic Shape Memory Alloys. *Adv. Eng. Mater.* **2012**, *14*, 547–561. [CrossRef]
34. Xu, N.; Raulot, J.M.; Li, Z.B.; Bai, J.; Yang, B.; Zhang, Y.D.; Meng, X.Y.; Zhao, X.; Zuo, L.; Esling, C. Composition-dependent structural and magnetic properties of Ni-Mn-Ga alloys studied by ab initio calculations. *J. Mater. Sci.* **2015**, *50*, 3825–3834. [CrossRef]
35. Dutta, B.; Çakır, A.; Giacobbe, C.; Al-Zubi, A.; Hickel, T.; Acet, M.; Neugebauer, J. *Ab initio* Prediction of Martensitic and Intermartensitic Phase Boundaries in Ni-Mn-Ga. *Phys. Rev. Lett.* **2016**, *116*, 025503. [CrossRef] [PubMed]
36. Sokolovskiy, V.V.; Gruner, M.E.; Entel, P.; Acet, M.; Çakır, A.; Baigutlin, D.R.; Buchelnikov, V.D. Segregation tendency of Heusler alloys. *Phys. Rev. Mater.* **2019**, *3*, 084413. [CrossRef]
37. Sokolovskaya, Y.A.; Sokolovskiy, V.V.; Zagrebin, M.A.; Buchelnikov, V.D.; Zayak, A.T. Ab Initio Study of the Composite Phase Diagram of Ni-Mn-Ga Shape Memory Alloys. *J. Exp. Theor. Phys.* **2017**, *125*, 104–110. [CrossRef]
38. Sokolovskiy, V.V.; Sokolovskaya, Y.A.; Zagrebin, M.A.; Buchelnikov, V.D.; Zayak, A.T. Ternary diagrams of magnetic properties of Ni-Mn-Ga Heusler alloys from *ab initio* and Monte Carlo studies. *J. Magn. Magn. Mater.* **2019**, *470*, 64–68. [CrossRef]
39. Sokolovskaya, Y.; Zagrebin, M.; Buchelnikov, V.; Zayak, A. Ternary phase diagram of Ni-Mn-Ga: Insights from ab initio calculations. *EPJ Web Conf.* **2018**, *185*, 05012. [CrossRef]
40. Ebert, H.; Koedderitzsch, D.; Minar, J. Calculating condensed matter properties using the KKR-Green's function method—Recent developments and applications. *Rep. Prog. Phys.* **2011**, *74*, 096501. [CrossRef]
41. Kresse, G.; Furthmüller, J. Efficient iterative schemes for ab initio total-energy calculations using a plane-wave basis set. *Phys. Rev. B* **1996**, *54*, 11169–11186. [CrossRef] [PubMed]
42. Kresse, G.; Joubert, D. From ultrasoft pseudopotentials to the projector augmented-wave method. *Phys. Rev. B* **1999**, *59*, 1758–1775. [CrossRef]
43. Anelli, A.; Engel, E.; Pickard, C.J.; Ceriotti, M. Generalized convex hull construction for materials discovery. *Phys. Rev. Mater.* **2018**, *2*, 103804-8. [CrossRef]
44. Perdew, J.P.; Burke, K.; Ernzerhof, M. Generalized Gradient Approximation Made Simple. *Phys. Rev. Lett.* **1996**, *77*, 3865–3868. [CrossRef] [PubMed]
45. Chernenko, V.A.; Cesari, E.; Kokorin, V.V.; Vitenko, I.N. The development of new ferromagnetic shape memory alloys in Ni-Mn-Ga system. *Scr. Metall. Mater.* **1995**, *33*, 1239–1244. [CrossRef]
46. Adachi, K.; Wayman, C.M. Transformation behavior of nearly stoichiometric Ni-Mn alloys. *Metall. Trans. A* **1985**, *16*, 1567–1579. [CrossRef]
47. Entel, P.; Dannenberg, A.; Siewert, M.; Herper, H.C.; Gruner, M.E.; Buchelnikov, V.D.; Chernenko, V.A. Composition-Dependent Basics of Smart Heusler Materials from First- Principles Calculations. *Mater. Sci. Forum* **2011**, *684*, 1–29. [CrossRef]

Article

Temperature Dependent Stress–Strain Behavior and Martensite Stabilization in Magnetic Shape Memory $Ni_{51.1}Fe_{16.4}Ga_{26.3}Co_{6.2}$ Single Crystal

Patricia Lázpita [1,*], Elena Villa [2], Francesca Villa [2] and Volodymyr Chernenko [1,3]

[1] Basque Center for Materials Applications and Nanostructures, University of Basque Country (UPV/EHU), Parque Científico UPV/EHU, 48940 Leioa, Spain; volodymyr.chernenko@ehu.eus
[2] CNR ICMATE Sede di Lecco, Via Previati n.1/e, 23900 Lecco, Italy; elena.villa@cnr.it (E.V.); francesca.villa@icmate.cnr.it (F.V.)
[3] Ikerbasque, Basque Foundation for Science, 48009 Bilbao, Spain
* Correspondence: patricia.lazpita@ehu.eus; Tel.: +34-94-601-5086

Citation: Lázpita, P.; Villa, E.; Villa, F.; Chernenko, V. Temperature Dependent Stress–Strain Behavior and Martensite Stabilization in Magnetic Shape Memory $Ni_{51.1}Fe_{16.4}Ga_{26.3}Co_{6.2}$ Single Crystal. *Metals* **2021**, *11*, 920. https://doi.org/10.3390/met11060920

Academic Editor: João Pedro Oliveira

Received: 27 April 2021
Accepted: 1 June 2021
Published: 4 June 2021

Publisher's Note: MDPI stays neutral with regard to jurisdictional claims in published maps and institutional affiliations.

Copyright: © 2021 by the authors. Licensee MDPI, Basel, Switzerland. This article is an open access article distributed under the terms and conditions of the Creative Commons Attribution (CC BY) license (https://creativecommons.org/licenses/by/4.0/).

Abstract: The superelastic properties and stress-induced martensite (SIM) stabilization have been studied in a shape memory $Ni_{51.1}Fe_{16.4}Ga_{26.3}Co_{6.2}$ single crystal. The single crystal, characterized by a thermally induced forward martensitic transformation temperature around 56 °C in the initial state, has been submitted to compression mechanical testing at different temperatures well above, near and below the martensitic transformation (MT). After each mechanical test, the characteristic MT temperatures and the transformation enthalpy have been monitored by means of differential scanning calorimetry. At temperatures below MT, the stress–strain (σ–ε) curves show a large strain, around 6.0%, resulting from the detwinning process in the martensitic microstructure, which remains accumulated after unloading in the detwinned state of the sample as a typical behavior of the shape memory alloys (SMAs). After just two "σ–ε + heating" cycles the accumulation of strain was not observed any more indicating the formation of a two-way shape memory effect which consists in a spontaneous recovery of the aforementioned detwinned state of the sample during its cooling across the forward MT. Whereas the thermally induced shape recovery in conventional SMAs occurs at the fixed value of the reverse MT temperature, the heating DSC curves of the mechanically deformed martensite in the present work show a burst-like calorimetric peak at the reverse MT arising at temperatures essentially higher than the thermally activated one. This behavior is the result of the SIM stabilization effect. After a short thermal aging in the stress-free state, this effect almost disappears, showing a slight impact on the MT characteristic temperatures and the enthalpy. At temperatures higher than the transformation one, the SIM is not stabilized, as the mechanically induced martensite fully retransforms into austenite after the unloading. From the σ–ε curves, the critical stress, σ_c, as well as the values of Young's moduli of martensite and austenite are determined showing linear dependences on the temperature with a slope of 3.6 MPa/°C.

Keywords: martensitic transformation; superelastic effect; stress-induced stabilization of martensite; critical stress; Young's modulus

1. Introduction

Ferromagnetic shape memory alloys (FSMAs) have become a subject of great interest for high technology applications due to the giant strains, up to 12%, that they can exhibit in response to the external mechanical stress or magnetic field [1–3]. The strong magnetoelastic coupling in these compounds allows controlling the large deformation in a reversible way by the applied magnetic field, thereby opening new possibilities in the development of actuators and sensors [4]. In addition to the shape memory effect (SME) and the superelastic effect (SE) showed by the conventional shape memory compounds, the FSMAs exhibit the so-called magnetic field-induced strain (MFIS). The main mechanism that governs this effect is the reorientation of martensitic variants, which requires an appreciable value of the equivalent magnetostress larger than the twinning stress [5]. Obviously, the stabilization of martensite can increase the twinning stress, thereby impeding a large MFIS.

Martensite stabilization can be achieved by thermal, chemical, mechanical or thermomechanical treatment (see, e.g., [6,7] and references therein). The different degrees of the martensite stabilization were obtained by heat treatment, e.g., in polycrystalline $Ni_{54}Fe_{16}Ga_{27}Mn_3$ and $Ni_{52}Fe_{18}Ga_{27}Co_3$ FSMAs, explained by a quenched-in disorder [8], or by aging of the stress-induced martensite (SIM-aging) in a single crystalline $Co_{49}Ni_{21}Ga_{30}$, explained within the concept of symmetry-conforming short-range order [9]. Recently, SIM-aging effects on the acoustic emission and the entropy accompanying MT were studied in the $Ni_{51}Fe_{18}Ga_{27}Co_4$ single crystal [10]. It is worth noting that in almost all aforementioned FSMAs the γ′-precipitates were present contributing, to some extent, to a martensite stabilization [8,9].

Stabilization of martensitic phase can also occur after its mechanical deformation resulting in the detwinned state of martensite, in this case it is called a stress-induced martensite stabilization (SIM stabilization). The fact of SIM stabilization, which is basically related to the formation of the non-equilibrium microstructure of martensitic phase, can be revealed by the enhanced temperature of the thermally induced reverse MT, whereas the forward MT remains intact. The difficulties in forming a habit plane between the austenite and the detwinned martensite were considered to be an origin of this mechanical stabilization effect [11,12].

The external mechanical stress can be replaced by the internal one that can also promote the nucleation and growth of the twin variants in a preferential direction [13]. The internal stress can be created through the formation of dislocations and other defects in the lattice, coherent particles, but also the SIM can promote internal stresses that induce a growth of the oriented variants. The formation of the SIM-induced effects, such as a two-way SME or rubber-like behavior of martensite, has been investigated in single crystals of CoNiGa, CoNiAl, NiMnGa and NiFeGaCo FSMAs [9,14–21] showing a strong dependency of the growth and stabilization of the martensitic variants on temperature, stress and crystallographic orientation. Therefore, the elucidation of the conditions responsible for the SIM stabilization in FSMAs is highly desirable since this stabilization directly impacts their functionalities that they exhibit as the magnetically activated materials or as conventional SMAs [9,10].

Among Heusler type FSMAs, the NiFeGaCo alloys represent (to date) the only known analog to classical NiMnGa in terms of a giant MFIS that they show [3,22]. Compared to NiMnGa, these materials are much more ductile (see [23] and references therein) whereby they are much more sustainable to SE cycling, which is an important advantage for their applications in elastocaloric devices [24–26]. These materials are also interesting for their magnetocaloric [27] and magnetoresistance properties [28].

In the present study, we examine the superelastic properties and the stress-induced martensite stabilization in a single crystal of NiFeGaCo FSMA with composition which is not prone to showing a second-phase precipitation [29,30]. The stress–strain measurements have been systematically performed on the sample cut along $<110>_A$ crystallographic axis of the cubic austenite since the deformation in this direction is crucial for obtaining a giant rubber-like behavior in an SIM-aged conditioned martensitic state [19], or developing a two-way SME [14], or modifying a character of the thermally induced strain recovery in such FSMAs [31]. Thus, we carried out the compression mechanical tests along the $<110>_A$ direction at different temperatures: well above, near and below the martensitic transformation, whereby assessing the temperature range of the SIM stabilization and the impact of the mechanical stress on the character of MT.

2. Materials and Methods

A single crystalline ingot with a composition of $Ni_{51.1}Fe_{16.4}Ga_{26.3}Co_{6.2}$ (at.%) was grown by a floating zone method, solution treated at 900 °C for 24 h and water quenched.

The <110>$_A$ crystallographic orientation in cubic austenitic phase of the single crystal was determined by the Laue back-reflection method and a sample with this orientation was cut using electro-discharge machine (EDM). The alloy composition was determined with an uncertainty of 0.5 at.% by energy dispersive X-ray spectrometry (EDXS) and characterized in the initial quenched state by a two-step reverse MT at heating and thermally induced MT temperature around 56 °C at cooling. The alloy composition and heat treatment ensured the formation of the non-modulated tetragonal 2M-martensitic phase [32,33]. The multi- or two-step character of the reverse MT was observed in the literature for different SMAs (see, e.g., [34,35]) and may be explained by the formation of the nonuniform microstructure of martensite resulting from quenching: presumably, the main body of the sample consists of a thermally stabilized one-twin variant co-existing with the rest multitwinned part of sample. Whereas the latter part transforms at a fixed reverse MT temperature, the former part exhibits a jerky-like MT at higher temperatures resulting in an extension of the entire interval of reverse MT.

A cylindrical sample of 3 mm diameter and about 4 mm height has been tested by compression along the crystallographic <110>$_A$ direction at different temperatures and by the differential scanning calorimetry (DSC). For the mechanical tests, we used a MTS 2/M mechanical testing machine in compression configuration with a load cell of 10 kN and a climatic chamber. In this configuration, we measured the σ–ε curves in a stress control mode up to 350 MPa with a loading/unloading rate of about 20 MPa/min at constant temperatures varying from room temperature to 130 °C. The compression conditions were selected to prevent a possible sample failure and to achieve better accuracy of the measurements. To perform each σ–ε experiment the sample was heated from the room temperature up to the temperature value of the following test. Before the test and after each σ–ε cycle, in order to determine the MT characteristics of the currently unloaded sample, we recorded its DSC heating and cooling curves with a TA Instruments Q100 Calorimeter, in the temperature range between 10 °C and 200 °C with a heating/cooling rate of 10 °C/min. As the free sample was always in a martensitic phase at room temperature, all DSC measurements were started by heating through the reverse MT until 200 °C, holding at this temperature for 10 min to allow the complete recovery of the sample deformation induced by the compression, and then cooling back to the room temperature.

The crystal structure has been checked by X-ray diffraction from the basement of the cylindrical sample with a plane of {110}$_A$ used for compression, by an X'pert Panalytical diffractometer at room temperature before the first mechanical test and after all the thermomechanical tests.

3. Results and Discussion

Figure 1 shows the compression stress–strain, σ–ε, curves along the <110>$_A$ direction of the single crystalline sample at selected temperatures below, in the interval and above the reverse MT temperature, A_s = 69 °C, which was determined in the initial state before the mechanical test (see Table 1). Each σ–ε curve was recorded after the thermal treatment of the free sample at 200 °C, so the latter was initially in the thermally induced martensitic phase ($T < A_s$), martensitic/austenite two-phase state ($A_s < T < A_f$) and in the austenitic phase ($T > A_f$). At temperatures below A_s, i.e., at 22 °C and 40 °C, the curves show a plateau corresponding to the irreversible twin variant reorientation process followed by the elastic deformation of the martensitic phase reaching maximum strain of 9% at 350 MPa at 22 °C. This induced strain is only partly recovered after the unloading resulting in a residual deformation of around 6% related to SIM stabilization. This strain value, resulting from the detwinning process in the martensitic microstructure, is in line with the obtained one in the <110>$_A$ compression tests of the martensitic phases in NiFeGaCo [19] and NiMnGa [36]. The curve at 60 °C, also obtained below A_s, does not show the strain accumulation, only indicates the presence of the quasi-elastic σ–ε loop almost identical to the one for the oriented martensite obtained at 40 °C test. This suggests a spontaneous recovery of the aforementioned detwinned state of the free sample during cooling through the forward MT.

Such behavior is obviously related to the two-way shape memory effect (TWSME) which was developed after the two "σ–ε + heating" cycles at 22 °C and 40 °C. TWSME implies a spontaneous macroscopic shape change in the sample induced by cooling/heating through MT, where the martensitic variants are oriented in a preferable direction under internal stresses produced by the defect structure formed via a thermomechanical training. Thus, we think, that namely TWSME is responsible for the absence of plateau at 60 °C. We assume that the previous two stages of the thermomechanical treatment were enough to "train" the sample whereby the same martensitic variant, memorized from the previous compression tests, is obtained after DSC cooling of the trained sample through the forward MT to the room temperature. Therefore, σ–ε test at 60 °C reveals an elastic behavior of the spontaneously oriented martensitic variant. TWSME was also observed for single crystalline NiMnGa [15,18], CoNiAl [14] and NiFeGaCo [21] FSMAs.

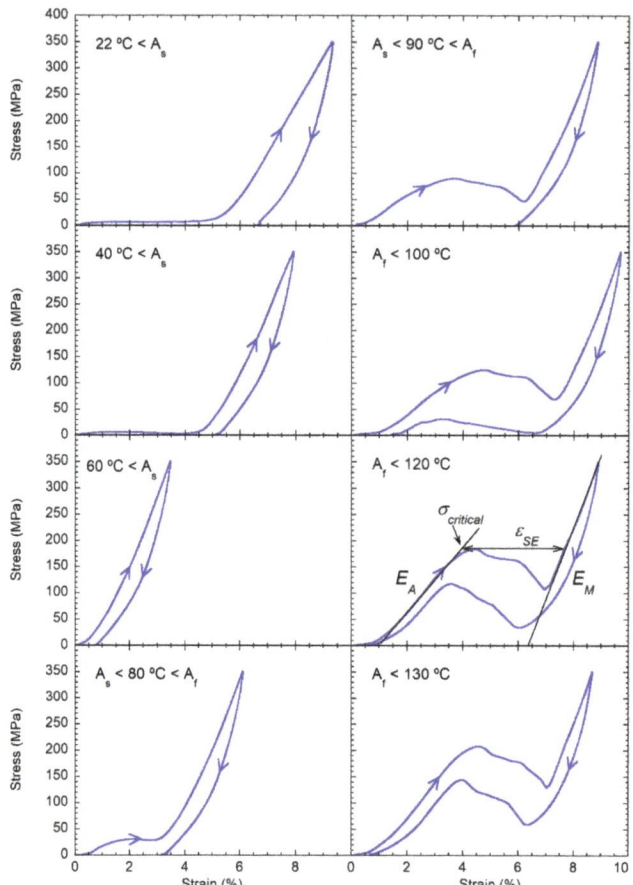

Figure 1. Stress–strain compression curves at different temperatures: below A_s = 69 °C, in the interval A_f-A_s, where A_f = 92 °C, and above A_f of the thermally induced reverse MT. The schematic at temperature 120 °C shows how the values of critical stress ($\sigma_{critical}$), superelastic strain (ε_{SE}), and Young's moduli of the austenitic (E_A) and martensitic (E_M) phases were determined.

Table 1. Characteristic parameters of the martensitic transformation before the compression tests and after the loading/unloading stress–strain (SS) cycles at different temperatures determined by means of the calorimetric measurements. The values of the forward martensitic transformation temperature start (M_s), finish (M_f) and enthalpy change ($\Delta H_{cooling}$) are obtained from the cooling ramps. Reverse martensitic transformation temperature start (A_s), finish (A_f) and the enthalpy change ($\Delta H_{heating}$) are extracted from the heating runs.

	Temperature of SS Test Cycle	M_s (°C)	M_f (°C)	$\Delta H_{cooling}$ (J/g)	A_s (°C)	A_f (°C)	$\Delta H_{heating}$ (J/g)
	before the tests	62	50	3.4	69	92	2.5
$A_s < T$	after SS 22 °C	61	48	3.1	101	103	3.1
	after SS 40 °C	59	47	3.0	102	103	3.1
	after SS 60 °C	59	47	3.0	101	102	3.0
$A_s < T < A_f$	after SS 80 °C	59	46	3.0	99	100	3.1
	after SS 90 °C	58	46	3.1	98	100	3.1
$T > A_f$	after SS 100 °C	59	46	3.2	65	77	3.3
	after SS 120 °C	58	47	3.4	62	76	3.1
	after SS 130 °C	58	47	3.4	63	77	3.0

At temperatures in the interval A_f-A_s and above A_f, the stress-induced MT and a conventional superelastic behavior are observed. The initial slope at low applied stress corresponds to the elastic deformation of the austenitic phase, and at the critical stress values the induced first portion of the martensitic phase manifests itself as a plateau-like maximum on the curves followed by a decrease in the stress. Once the stress-induced MT is completed, the elastic deformation of the martensite is observed at high stresses.

At 80 °C and 90 °C, SE is incomplete due to the existence of thermal hysteresis of MT, so the unrecoverable residual strain remains after the unloading. This effect relates to the austenite/martensite mixture at these temperatures, hence only an austenite fraction is transforming.

At temperatures above A_f, the induced strain is perfectly superelastic reflecting a complete stress-induced MT from the austenitic phase to the martensitic one by loading that is fully reversible after unloading. In this case, there is no residual deformation after σ–ε cycle.

In all experiments in Figure 1, where the stress-induced MT is observed, the curves show a non-monotone stress–strain dependence just after achieving the critical stress necessary for the transformation. Such non-monotone σ–ε dependencies of the forward and reverse stress-induced MT indicate the non-equilibrium progression of the transformation between the austenitic and martensitic phases and is linked to the different stresses required for the nucleation and propagation of the interfaces between phases [37]. Actually, in [31] this compression behavior for the <110>$_A$ oriented NiFeGaCo single crystal was modeled in terms of the direct relationship between the elastic interface stresses and the MT kinetics. Similar tendencies have been also observed in the experiments on the NiMnGa, CoNiGa and CoNiAl single crystals [17,20,37,38] which show a non-monotonous behavior for applied stress in the <110> direction, as occurs in our specimen.

The elastic branches in the loading curves for the martensitic phase in Figure 1 show a rather linear behavior, whereas unloading branches demonstrate nonlinearity meaning that the sample expansion response does not follow the rate of stress removal. Therefore, the hysteretic behavior in this elastic branch is considered as the mostly instrumental one which may be attributed to the influence of frictional forces between the anvils and sample.

The data in Figure 1 have been used for the evaluation of the different parameters of the stress-induced MT according to the schematic depicted in the same figure at 120 °C. The total strain achieved in the superelastic region, ε_{SE}, was obtained as the difference between the strain in the austenitic phase and the martensitic one at the critical stress point (see schematic in Figure 1). One of the results is a moderate decrement of ε_{SE} with the temperature, from 4.2% at 100 °C to 3.6% at 130 °C, that may be associated with the growth of multiple variants due to improvement of the variant–variant interaction at higher temperatures [39]. The stress hysteresis of the MT remains practically unchanged in the studied temperature range.

From stress–strain curves we have determined the phase diagram "critical stress versus temperature", $\sigma_{critical}$ (T), and the Young's moduli of the austenite (E_A) and martensite (E_M). The values of the critical stress or twinning stress at each temperature were defined by the two-tangent method, as the cross point of the linearly extrapolated slope related to the elasticity of either austenite or martensite and the extended straight line of the plateau region (see schematic in Figure 1). Figure 2a displays the twinning stress required for the variant reorientation in the martensitic phase at $T < A_s$ (open squares) and the critical stress of the stress-induced MT at $T > A_f$ (solid circles). It is seen that, while the twinning stress is slightly decreasing with the temperature, the critical stress for stress-induced MT linearly increases as a function of the temperature with a slope of 3.6 MPa/°C. This value is in line with those found in other Ni-Fe-Ga-based FSMAs [33,40]. It is easy to find that the extrapolated temperature value at zero stress, which gives the transformation temperature from the thermally induced martensite to the austenite, is in a good agreement with the DSC results (69 °C) (see Table 1). Figure 2b depicts temperature dependences of the Young's moduli of the austenite, E_A, and martensitic phase, E_M, evaluated from the slopes of the stress–strain curves corresponding to the quasilinear elastic deformation in the loading process. It has to be noted that E_M is much higher than E_A and presents two different dependencies on temperature, 100 MPa/°C for temperatures below MT, i.e., for the thermally induced martensite, and 34 MPa/°C at high temperatures for the stress-induced martensite. These tendencies are indicative of the different martensitic microstructures formed as a result of the free-sample cooling through MT (poly-variant martensitic phase) or as a result of the stress-induced MT (the sample consists of mainly one-variant of martensite). On the other hand, E_A presents a smooth increment with the temperature having a slope of 85 MPa/°C close in value to the temperature dependence of poly-variant state of martensite. The $E_A(T)$ behavior agrees with the one observed in other FSMAs, such as, e.g., Co-Ni-Al or Ni-Mn-Ga, and has been explained by the decrement of the elastic constants due to the lattice softening when approaching the start of the forward MT [17,41].

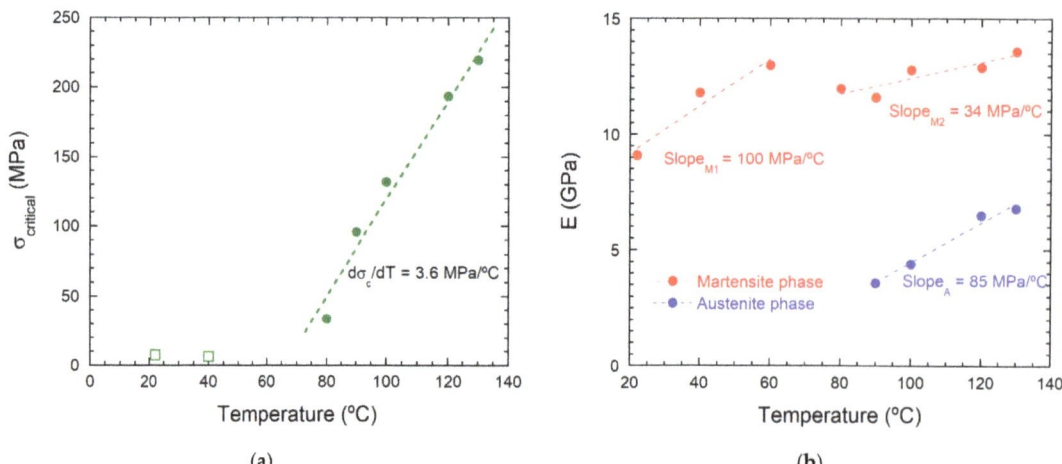

Figure 2. (a) Temperature dependence of the critical stress of the MT start (solid circles) and twinning stress (open squares). (b) Evolutions of Young's moduli, E_A and E_M, with the temperature. All lines are the linear fits to the data points.

To check the impact of the SIM on MT, after each mechanical test the characteristic MT temperatures of the sample and the enthalpy of the transformation (ΔH) have been monitored by means of differential scanning calorimetry. To this aim, the DSC curves were recorded during heating the sample to 200 °C, holding for 10 min and subsequent cooling to the room temperature. The results of DSC measurements are shown in Figure 3 and Table 1. The anomalous shapes on DSC curves in Figure 3 vary strongly when one compares exothermic and some endothermic peaks. Whereas the former ones have a common cupola shape, the latter ones exhibit a burst-like behavior characterized by a very fast heat absorption causing a loop on the curve due to a short overcooling of the sample. In the case of such abnormal loops, the values of ΔH were determined by the integration of DSC signal over a time.

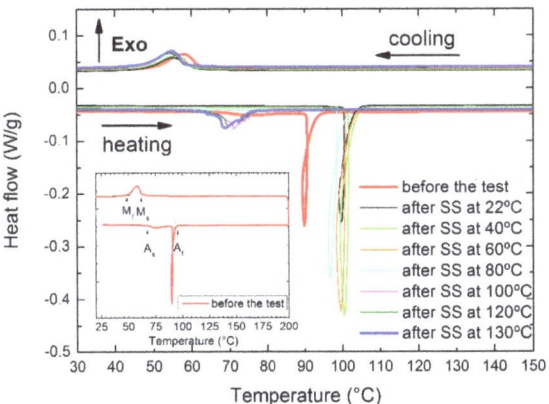

Figure 3. DSC heating and cooling runs measured before the first stress–strain (SS) test and after each compression SS cycle at the constant temperature indicated in the graphs. The inset illustrates the determination of the MT characteristic temperatures in the as-received sample before testing for which the two-step reverse MT is observed on the heating curve (see experimental section).

It is instructive to note that thermally activated forward MT during cooling gives rise to the overlapping single broad peaks. Thus, the different mechanical tests show a low impact on the forward MT only slightly affecting both the MT enthalpy, $\Delta H_{cooling} = (3.2 \pm 0.2)$ J/g, as the average value, and the MT start temperature, M_s, that varies in a range between 62 °C and 58 °C.

The DSC heating runs in Figure 3 reveal the different influence of the mechanical tests on the thermally activated reverse MT if they are compared with the cooling runs. The reverse MT manifests in DSC curves through two different behaviors. Firstly, after mechanical tests at temperatures lower than 100 °C ($A_f < 100$ °C), which involve a primary deformation of martensitic phase or its fraction, DSC heating runs show a sharp peak associated to the burst-like behavior in a range of temperatures between 98 °C and 102 °C shifting towards high temperatures, around 30 °C (Table 1), due to the effect of stress-induced stabilization of the martensite. Table 1 shows that the average values of $\Delta H_{heating}$ and $\Delta H_{cooling}$ are almost the same indicating the same amount of martensite involved in the thermally induced MT independently of mechanical tests. The reduced value of enthalpy for the reverse MT in the as-received sample, calculated using areas under two DSC peaks (Figure 3), could be due to an underestimation produced by the difficulty of area estimation for the very smeared first peak. After a short thermal aging of the stress-free sample at 200 °C, the DSC cooling runs exhibit peaks with characteristics corresponding to the initial state of the sample.

Secondly, after the mechanical tests at 100 °C and above, where the SE is complete, i.e., the strain is fully recovered due to the transformation into austenite, a broad DSC peak characterized by $A_s = 65$ °C (close to $A_s = 69$ °C of the diffuse peak for the as-received sample, Table 1) is observed, indicating no SIM stabilization effect. Therefore, it is clear that the stress-induced reorientation of the martensitic variants promotes the stabilization of the martensitic phase to the higher temperatures by increasing A_s from (63–69) °C to (98–102) °C (Table 1). The difficulties in forming a habit plane between austenite and the twinned/detwinned martensite, due to their lattice incompatibility, are considered to be an origin of the SIM stabilization effect (see, e.g., [11,12,35]).

Furthermore, X-ray diffraction patterns were measured with the intention to throw some light on the effect of the compression/temperature treatment on the growth of different martensitic variants. Diffractograms shown in Figure 4 were obtained at room temperature before the mechanical test, in the initial state of sample, and after all the "σ–ε + heating" cycles. The single crystal was oriented along the <110>$_A$ axis, so before the compression two reflections that correspond to the differently oriented martensitic variants are observed. These reflections fit with the non-modulated tetragonal unit cell with lattice parameters of a = 0.542 nm and c = 0.648 nm. This tetragonal unit cell is considered in the coordinates of the austenitic L2$_1$-ordered cubic lattice, whereby during MT it shrinks along two axes and expands along the third one. In this case, the value of c/a ratio is equal to about 1.2 which is a common value for FSMAs exhibiting a non-modulated tetragonal martensitic structure [19]. The cell parameters can be used to calculate the maximum possible SE strain which could be obtained due to a stress-induced MT. For that, the cell parameter of the cubic phase was accepted to be a_0 = 0.588 nm which was measured for Ni$_{51}$Fe$_{18}$Ga$_{27}$Co$_4$ [19,42]. Taking into account the above values of a, c and a_0, in compression tests along <110>$_A$ direction the maximum calculated strain is equal to about 6.0% which is by about 40% larger than the experimental strains ε_{SE} determined from the experimental data in Figure 1 by the method shown inside it at T = 120 °C. This difference is indicative of the multivariant martensitic state induced during compressions tests, which is even more relevant at higher temperatures.

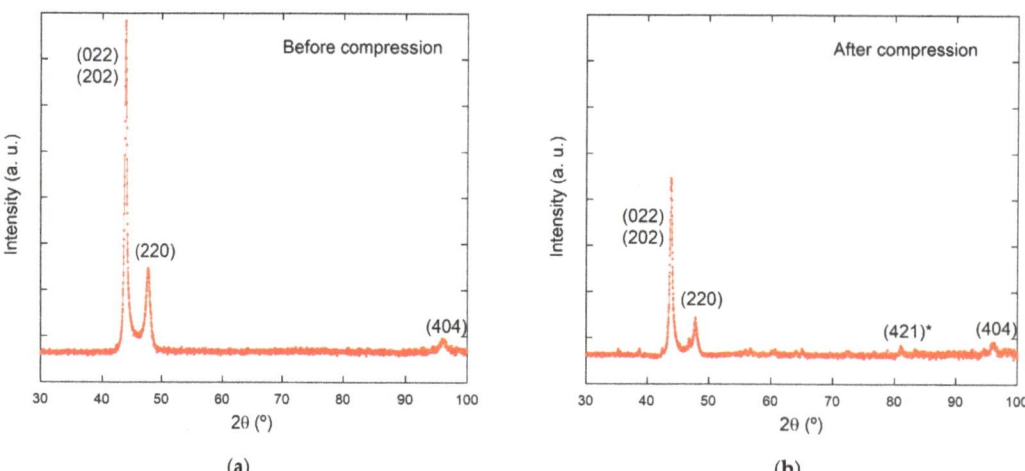

Figure 4. X-ray diffraction patterns of the basement surface of the sample in the martensitic state measured at room temperature: (a) before thermomechanical testing and (b) after the thermomechanical treatment.

It has to be noted that after compression in the <110>$_A$ direction, the main reflections remain at the same position, but their intensities are drastically reduced alongside the appearance of a new tiny peak indicated in Figure 4 by an asterisk. All these features in diffraction patterns could be related to the growth or disappearance of differently oriented martensitic variants. In order to determine the redistribution of the different martensitic variants, we estimate the relative change in the intensity between different reflections. To this aim we calculate the ratio between the integrated intensity of the (202)/(022) and (220) reflections before and after compression. Before the compression, there is a random variant orientation of one-third for each direction, with the relation 1:1:1 for the reflections (022), (202) and (220), respectively. However, after the cycling there is a drastic intensity reduction, by half, of the intensity of the principal reflections, although they present practically the same distribution ratio of 1:1:1. This fact can be related to the growth of variants with other orientations that would be favored by residual internal stresses or defects generated during the thermomechanical treatment. Moreover, in the diffraction pattern there are no peaks corresponding to the austenite phase, which corroborates the complete reversibility of the stress-induced MT confirming the DSC results.

4. Summary

The stress–strain behavior and stress-induced martensite (SIM) stabilization have been studied in a shape memory $Ni_{51.1}Fe_{16.4}Ga_{26.3}Co_{6.2}$ (at.%) single crystal under compression along the <110>$_A$ direction at different temperatures: well above, near and below the reverse martensitic transformation. At temperatures below MT, the stress–strain curves show a large strain, around 6.0%, related to the detwinning process in the martensitic microstructure, which keeps accumulating after unloading due to the stabilized character of the mechanically deformed martensite. Calorimetric analysis shows a burst-like calorimetric peak at about 100 °C which corresponds to the reverse MT of mechanically deformed martensite. This value of the reverse MT start temperature is much higher than the thermally activated one (69 °C) obtained for the sample in its initial state, reflecting a martensitic phase stabilization. The difficulties in forming a habit plane between the austenite and the twinned/detwinned martensite, due to their lattice incompatibility, are considered to be an origin of the SIM stabilization effect. At temperatures from 100 °C to 130 °C, the superelastic effect is entirely reversible. The superelastic strain was found to be reduced, if compared with the one calculated using the lattice parameters, due to the growth of multiple variants. In contrast to mechanically deformed martensite, SIM at high temperatures, induced from austenitic phase, is not stabilized due to a complete recovery after the stress unloading through reverse MT. The critical stress, σ_c, versus temperature phase diagram shows a linear increment when the temperature increases, with a slope $d\sigma_c/dT$ of 3.6 MPa/°C. Moreover, in this case the characteristic temperatures of the thermally induced MT remain practically constant after the stress cycling.

The present work demonstrates that the phenomenon of SIM stabilization should be taken into account in the case of the need for cyclic actuation using a single crystalline FSMA.

Author Contributions: Conceptualization, E.V. and V.C.; data curation, P.L., E.V. and F.V.; formal analysis, P.L., E.V. and F.V.; funding acquisition, E.V. and V.C.; investigation, P.L., E.V. and F.V.; methodology, E.V. and F.V.; resources, E.V. and F.V.; supervision, E.V. and V.C.; writing—original draft, P.L. and V.C.; writing—review and editing, P.L., E.V. and V.C. All authors have read and agreed to the published version of the manuscript.

Funding: This research has been carried out with the financial support of the Spanish Ministry of Science, Innovation and Universities (project RTI2018-094683-B-C53-54) and Basque Government Department of Education (project IT1245-19) and in the framework of INNOSMAD ID546749 Project, PROGRAMMA DI COOPERAZIONE INTERREG V-A ITALIA SVIZZERA CCI 2014TC16RFCB035.

Conflicts of Interest: The authors declare no conflict of interest.

References

1. Sozinov, A.; Lanska, N.; Soroka, A.; Zou, W. 12% magnetic field-induced strain in Ni-Mn-Ga-based non-modulated martensite. *Appl. Phys. Lett.* **2013**, *102*, 021902. [CrossRef]
2. Müllner, P.; Chernenko, V.A.; Kostorz, G. Large cyclic magnetic-field-induced deformation in orthorhombic (14M) Ni–Mn–Ga martensite. *J. Appl. Phys.* **2004**, *95*, 1531–1536. [CrossRef]
3. Morito, H.; Fujita, A.; Oikawa, K.; Ishida, K.; Fukamichi, K.; Kainuma, R. Stress-assisted magnetic-field-induced strain in Ni–Fe–Ga–Co ferromagnetic shape memory alloys. *Appl. Phys. Lett.* **2007**, *90*, 062505. [CrossRef]
4. Pagounis, E.; Muellner, P. Materials and Actuator Solutions for Advanced Magnetic Shape Memory Devices. In Proceedings of the ACTUATOR 2018, 16th International Conference on New Actuators, Bremen, Germany, 25–27 June 2018; pp. 1–7.

5. Heczko, O. Magnetic shape memory effect and highly mobile twin boundaries. *Mater. Sci. Technol. UK* **2014**, *30*, 1559–1578. [CrossRef]
6. Kustov, S.; Pons, J.; Cesari, E.; Van Humbeeck, J. Chemical and mechanical stabilization of martensite. *Acta Mater.* **2004**, *52*, 4547–4559. [CrossRef]
7. Oliveira, J.P.; Fernandes, F.M.B.; Schell, N.; Miranda, R.M. Martensite stabilization during superelastic cycling of laser welded NiTi plates. *Mater. Lett.* **2016**, *171*, 273–276. [CrossRef]
8. Picornell, C.; Pons, J.; Cesari, E.; Dutkiewicz, J. Thermal characteristics of Ni–Fe–Ga–Mn and Ni–Fe–Ga–Co ferromagnetic shape memory alloys. *Intermetallics* **2008**, *16*, 751–757. [CrossRef]
9. Niendorf, T.; Krooß, P.; Somsen, C.; Eggeler, G.; Chumlyakov, Y.I.; Maier, H.J. Martensite aging-Avenue to new high temperature shape memory alloys. *Acta Mater.* **2015**, *89*, 298–304. [CrossRef]
10. Samy, N.M.; Daróczi, L.; Tóth, L.Z.; Panchenko, E.; Chumlyakov, Y.; Surikov, N.; Beke, D.L. Effect of stress-induced martensite stabilization on acoustic emission characteristics and the entropy of martensitic transformation in shape memory $Ni_{51}Fe_{18}Ga_{27}Co_4$ single crystal. *Metals* **2020**, *10*, 534. [CrossRef]
11. Picornell, C.; Pons, J.; Cesari, E. Stabilisation of martensite by applying compressive stress in Cu-Al-Ni single crystals. *Acta Mater.* **2001**, *49*, 4221–4230. [CrossRef]
12. Roytburd, A.L. Intrinsic hysteresis of superelastic deformation. *Mater. Sci. Forum* **2000**, *327–328*, 389–392. [CrossRef]
13. Singh, S.; Kushwaha, P.; Scheibel, F.; Liermann, H.P.; Barman, S.R.; Acet, M.; Felser, C.; Pandey, D. Residual stress induced stabilization of martensite phase and its effect on the magnetostructural transition in Mn-rich Ni-Mn-In/Ga magnetic shape-memory alloys. *Phys. Rev. B-Condens. Matter Mater. Phys.* **2015**, *92*, 1–6. [CrossRef]
14. Panchenko, E.; Eftifeeva, A.; Chumlyakov, Y.; Gerstein, G.; Maier, H.J. Two-way shape memory effect and thermal cycling stability in $Co_{35}Ni_{35}Al_{30}$ single crystals by low-temperature martensite ageing. *Scr. Mater.* **2018**, *150*, 18–21. [CrossRef]
15. Timofeeva, E.E.; Panchenko, E.Y.; Pichkaleva, M.V.; Tagiltsev, A.I.; Chumlyakov, Y.I. The effect of stress-induced martensite ageing on the two-way shape memory effect in $Ni_{53}Mn_{25}Ga_{22}$ single crystals. *Mater. Lett.* **2018**, *228*, 490–492. [CrossRef]
16. Panchenko, E.; Chumlyakov, Y.; Maier, H.J.; Timofeeva, E.; Karaman, I. Tension/compression asymmetry of functional properties in [001]-oriented ferromagnetic NiFeGaCo single crystals. *Intermetallics* **2010**, *18*, 2458–2463. [CrossRef]
17. Li, P.; Karaca, H.E.; Chumlyakov, Y.I. Orientation dependent compression behavior of $Co_{35}Ni_{35}Al_{30}$ single crystals. *J. Alloys Compd.* **2017**, *718*, 326–334. [CrossRef]
18. Chernenko, V.A.; Villa, E.; Besseghini, S.; Barandiarán, J.M. Giant two-way shape memory effect in high-temperature Ni–Mn–Ga single crystal. *Phys. Procedia* **2010**, *10*, 94–98. [CrossRef]
19. Panchenko, E.; Timofeeva, E.; Eftifeeva, A.; Osipovich, K.; Surikov, N.; Chumlyakov, Y.; Gerstein, G.; Maier, H.J. Giant rubber-like behavior induced by martensite aging in $Ni_{51}Fe_{18}Ga_{27}Co_4$ single crystals. *Scr. Mater.* **2019**, *162*, 387–390. [CrossRef]
20. Chernenko, V.A.; Pons, J.; Cesari, E.; Zasimchuk, I.K. Transformation behaviour and martensite stabilization in the ferromagnetic Co–Ni–Ga Heusler alloy. *Scr. Mater.* **2004**, *50*, 225–229. [CrossRef]
21. Masdeu, F.; Pons, J.; Chumlyakov, Y.; Cesari, E. Two-way shape memory effect in $Ni_{49}Fe_{18}Ga_{27}Co_6$ ferromagnetic shape memory single crystals. *Mater. Sci. Eng. A* **2021**, *805*, 140543. [CrossRef]
22. Masdeu, F.; Pons, J.; Cesari, E.; Kustov, S.; Chumlyakov, Y.I. Magnetic-field-induced strain assisted by tensile stress in $L1_0$ martensite of a Ni–Fe–Ga–Co alloy. *Appl. Phys. Lett.* **2008**, *93*, 152503. [CrossRef]
23. Biswas, A.; Singh, G.; Sarkar, S.K.; Krishnan, M.; Ramamurty, U. Hot deformation behavior of Ni-Fe-Ga-based ferromagnetic shape memory alloy-A study using processing map. *Intermetallics* **2014**, *54*, 69–78. [CrossRef]
24. Pataky, G.J.; Ertekin, E.; Sehitoglu, H. Elastocaloric cooling potential of NiTi, Ni_2FeGa, and CoNiAl. *Acta Mater.* **2015**, *96*, 420–427. [CrossRef]
25. Xiao, F.; Jin, M.; Liu, J.; Jin, X. Elastocaloric effect in $Ni_{50}Fe_{19}Ga_{27}Co_4$ single crystals. *Acta Mater.* **2015**, *96*, 292–300. [CrossRef]
26. Li, Y.; Zhao, D.; Liu, J. Giant and reversible room-temperature elastocaloric effect in a single-crystalline Ni-Fe-Ga magnetic shape memory alloy. *Sci. Rep.* **2016**, *6*, 1–11. [CrossRef]
27. Sarkar, S.K.; Biswas, A.; Babu, P.D.; Kaushik, S.D.; Srivastava, A.; Siruguri, V.; Krishnan, M. Effect of partial substitution of Fe by Mn in $Ni_{55}Fe_{19}Ga_{26}$ on its microstructure and magnetic properties. *J. Alloys Compd.* **2014**, *586*, 515–523. [CrossRef]
28. Barandiarán, J.M.; Chernenko, V.A.; Lázpita, P.; Gutiérrez, J.; Feuchtwanger, J. Effect of martensitic transformation and magnetic field on transport properties of Ni-Mn-Ga and Ni-Fe-Ga Heusler alloys. *Phys. Rev. B* **2009**, *80*, 104404. [CrossRef]
29. Liu, J.; Scheerbaum, N.; Hinz, D.; Gutfleisch, O. Martensitic transformation and magnetic properties in Ni–Fe–Ga–Co magnetic shape memory alloys. *Acta Mater.* **2008**, *56*, 3177–3186. [CrossRef]
30. Sofronie, M.; Tolea, F.; Kuncser, V.; Valeanu, M. Martensitic transformation and accompanying magnetic changes in Ni-Fe-Ga-Co alloys. *J. Appl. Phys.* **2010**, *107*, 1–6. [CrossRef]
31. Nikolaev, V.I.; Yakushev, P.N.; Malygin, G.A.; Averkin, A.I.; Pulnev, S.A.; Zograf, G.P.; Kustov, S.B.; Chumlyakov, Y.I. Influence of partial shape memory deformation on the burst character of its recovery in heated Ni–Fe–Ga–Co alloy crystals. *Tech. Phys. Lett.* **2016**, *42*, 399–402. [CrossRef]
32. Oikawa, K.; Saito, R.; Anzai, K.; Ishikawa, H.; Sutou, Y.; Omori, T.; Yoshikawa, A.; Chernenko, V.A.; Besseghini, S.; Gambardella, A.; et al. Elastic and Superelastic Properties of NiFeCoGa Fibers Grown by Micro-Pulling-Down Method. *Mater. Trans.* **2009**, *50*, 934–937. [CrossRef]

33. Kosogor, A.; L'vov, V.A.; Chernenko, V.A.; Villa, E.; Barandiaran, J.M.; Fukuda, T.; Terai, T.; Kakeshita, T. Hysteretic and anhysteretic tensile stress-strain behavior of Ni-Fe(Co)-Ga single crystal: Experiment and theory. *Acta Mater.* **2014**, *66*, 79–85. [CrossRef]
34. Liu, N.; Huang, W.M. DSC study on temperature memory effect of NiTi shape memory alloy. *Trans. Nonferrous Met. Soc. China (Engl. Ed.)* **2006**, *16*, s37–s41. [CrossRef]
35. Hamilton, R.; Efstathiou, C.; Sehitoglu, H.; Chumlyakov, Y. Thermal and stress-induced martensitic transformations in NiFeGa single crystals under tension and compression. *Scr. Mater.* **2006**, *54*, 465–469. [CrossRef]
36. Panchenko, E.; Timofeeva, E.; Pichkaleva, M.; Tokhmetova, A.; Surikov, N.; Tagiltsev, A.; Chumlyakov, Y. Effect of Stress-Induced Martensite Aging on Martensite Variant Reorientation Strain in NiMnGa Single Crystals. *Shape Mem. Superelasticity* **2020**, *6*, 29–34. [CrossRef]
37. L'vov, V.A.; Rudenko, A.A.; Chernenko, V.A.; Cesari, E.; Pons, J.; Kanomata, T. Stress-induced martensitic transformation and superelasticity of alloys: Experiment and theory. *Mater. Trans.* **2005**, *46*, 790–797. [CrossRef]
38. Karaca, H.E.; Karaman, I.; Lagoudas, D.C.; Maier, H.J.; Chumlyakov, Y.I. Recoverable stress-induced martensitic transformation in a ferromagnetic CoNiAl alloy. *Scr. Mater.* **2003**, *49*, 831–836. [CrossRef]
39. Dadda, J.; Maier, H.J.; Karaman, I.; Karaca, H.E.; Chumlyakov, Y.I. Pseudoelasticity at elevated temperatures in [001] oriented $Co_{49}Ni_{21}Ga_{30}$ single crystals under compression. *Scr. Mater.* **2006**, *55*, 663–666. [CrossRef]
40. Villa, E.; Agilar-Ortiz, C.O.; Álvarez-Alonso, P.; Camarillo, J.P.; Lara-Rodriguez, G.A.; Flores-Zúñiga, H.; Chernenko, V.A. Shape memory behavior of Ni-Fe-Ga and Ni-Mn-Sn ribbons. *MATEC Web Conf.* **2015**, *33*. [CrossRef]
41. Zhao, P.; Dai, L.; Cullen, J.; Wuttig, M. Magnetic and Elastic Properties of $Ni_{49.0}Mn_{23.5}Ga_{27.5}$ Premartensite. *Metall. Mater. Trans. A* **2007**, *38*, 745–751. [CrossRef]
42. Oikawa, K.; Omori, T.; Sutou, Y.; Morito, H.; Kainuma, R.; Ishida, K. Phase Equilibria and Phase Transition of the Ni–Fe–Ga Ferromagnetic Shape Memory Alloy System. *Metall. Mater. Trans. A* **2007**, *38*, 767–776. [CrossRef]

Article

External-Field-Induced Phase Transformation and Associated Properties in a $Ni_{50}Mn_{34}Fe_3In_{13}$ Metamagnetic Shape Memory Wire

Zhen Chen [1], Daoyong Cong [1,*], Shilei Li [1], Yin Zhang [1], Shaohui Li [1], Yuxian Cao [1], Shengwei Li [1], Chao Song [1], Yang Ren [2] and Yandong Wang [1]

1. Beijing Advanced Innovation Center for Materials Genome Engineering, State Key Laboratory for Advanced Metals and Materials, University of Science and Technology Beijing, Beijing 100083, China; b20160442@xs.ustb.edu.cn (Z.C.); lishilei@ustb.edu.cn (S.L.); yinzhang330@163.com (Y.Z.); b20160443@xs.ustb.edu.cn (S.L.); b20190501@xs.ustb.edu.cn (Y.C.); b20200522@xs.ustb.edu.cn (S.L.); b20180487@xs.ustb.edu.cn (C.S.); ydwang@ustb.edu.cn (Y.W.)
2. X-ray Science Division, Argonne National Laboratory, Argonne, IL 60439, USA; ren@aps.anl.gov
* Correspondence: dycong@ustb.edu.cn; Tel.: +86-10-6233-2508

Citation: Chen, Z.; Cong, D.; Li, S.; Zhang, Y.; Li, S.; Cao, Y.; Li, S.; Song, C.; Ren, Y.; Wang, Y. External-Field-Induced Phase Transformation and Associated Properties in a $Ni_{50}Mn_{34}Fe_3In_{13}$ Metamagnetic Shape Memory Wire. Metals 2021, 11, 309. https://doi.org/10.3390/met11020309

Academic Editor: Ryosuke Kainuma

Received: 14 January 2021
Accepted: 5 February 2021
Published: 10 February 2021

Publisher's Note: MDPI stays neutral with regard to jurisdictional claims in published maps and institutional affiliations.

Copyright: © 2021 by the authors. Licensee MDPI, Basel, Switzerland. This article is an open access article distributed under the terms and conditions of the Creative Commons Attribution (CC BY) license (https://creativecommons.org/licenses/by/4.0/).

Abstract: Metamagnetic shape memory alloys exhibit a series of intriguing multifunctional properties and have great potential for applications in magnetic actuation, sensing and magnetic refrigeration. However, the poor mechanical properties of these alloys with hardly any tensile deformability seriously limit their practical application. In the present work, we developed a Ni-Fe-Mn-In microwire that exhibits both a giant, tensile superelasticity and a magnetic-field-induced first-order phase transformation. The recoverable strain of superelasticity is more than 20% in the temperature range of 233–283 K, which is the highest recoverable strain reported heretofore in Ni-Mn-based shape memory alloys (SMAs). Moreover, the present microwire exhibits a large shape memory effect with a recoverable strain of up to 13.9% under the constant tensile stress of 225 MPa. As a result of the magnetic-field-induced first-order phase transformation, a large reversible magnetocaloric effect with an isothermal entropy change ΔS_m of 15.1 J kg^{-1} K^{-1} for a field change from 0.2 T to 5 T was achieved in this microwire. The realization of both magnetic-field and tensile-stress-induced transformations confers on this microwire great potential for application in miniature multi-functional devices and provides an opportunity for multi-functional property optimization under coupled multiple fields.

Keywords: metamagnetic shape memory alloy; microwire; superelasticity; martensitic transformation; magnetocaloric effect; magnetic-field-induced phase transformation; magnetostructural transformation; shape memory effect

1. Introduction

Shape memory alloys (SMAs), as a unique class of smart materials, which combine the functional properties such as shape memory effect (SME), superelasticity (SE) and elastocaloric effect, have drawn great attention in recent years [1,2]. The underlying mechanism of these properties is a reversible phase transformation between a high-temperature austenitic phase and a low-temperature martensitic phase when an external stimulus of stress or temperature is applied [3]. Thus, SMAs show great potential for application as actuators and sensors in industrial [4,5], automotive [2,6], aerospace [3,7], micro-electromechanical system (MEMS) [8] and biomedical [9] fields. For the majority of actuators and sensors, fast response and large output strain/stress are imperative and desirable properties. Metamagnetic shape memory alloys (MMSMAs) have provoked much interest in recent years due to their high response frequency and output stress arising from the strong coupling of crystal and magnetic structures. In these alloys, it is possible to obtain such multifunctional properties as magnetic superelasticity [10,11], magnetic shape memory effect [12], magnetoresistance [13] and magnetocaloric effect [14–16].

In principle, MMSMAs can display a phase transformation under an external field of magnetic field, stress or temperature and optimized multifunctional properties can be anticipated under the coupling of magnetic field, stress and temperature. Unfortunately, polycrystalline MMSMAs show intrinsic brittleness, as a result of deformation and transformation incompatibility at grain boundaries and triple junctions [17], which severely limits their practical application. Moreover, it is difficult for these alloys to serve under the condition of the simultaneous application of magnetic field and stress in order to exhibit optimized multifunctional properties because they can easily fracture under external stress due to their high brittleness. It is of great importance to develop MMSMAs with good mechanical properties so that they can bear a high enough stress for stress-induced martensitic transformation.

Recently, it was proposed that the deformation and transformation incompatibility at grain boundaries and triple junctions in SMAs could be diminished by reducing their dimensions [18]. Low-dimension SMAs (particles, wires, films, ribbons, micropillars or foams) exhibit great application potential in micro-actuators or micro-sensors. This is due to the high ratio of surface to volume, which could improve the response speed [8,19]. The Taylor–Ulitovsky [20,21] and melt-extraction [22] methods are two feasible and easy methods to produce magnetic shape memory microwires. The Taylor–Ulitovsky method, which involves rapid solidification and drawing, is prone to produce microwires with an oligocrystalline structure. This structure reduces the incompatibility between different grains and thus effectively enhances the mechanical properties of Ni-Mn-based MMSMAs [23,24]. Although scattered attempts have been made to investigate the external-field-induced transformation of Ni-Mn-based MMSMA microwires that were prepared using this method, the reported microwires with a magnetic-field-induced transformation exhibit a limited recoverable strain of superelasticity and those with a large recoverable strain barely display magnetic-field-induced transformation. Therefore, there is an urgent need to develop high-performance MMSMAs with both a giant recoverable strain and a magnetic-field-induced first-order phase transition.

In the present work, we successfully developed a Ni-Mn-Fe-In MMSMA microwire exhibiting a giant, tensile superelasticity with a recoverable strain higher than 20%. Furthermore, the microwire shows a magnetic-field-induced first-order phase transformation and a reversible isothermal magnetic entropy change of 15.1 J kg^{-1} K^{-1} for a field change from 0.2 T to 5 T. The simultaneous achievement of a magnetic-field-induced phase transformation and a giant, tensile recoverable strain confers on this microwire great potential for application in miniature multifunctional devices.

2. Materials and Methods

$Ni_{50}Mn_{34}Fe_3In_{13}$ (at. %) polycrystalline button ingots were prepared by arc melting the pure Ni, Mn, Fe, and In elements. The ingots were melted four times in order to ensure homogeneity. The Taylor–Ulitovsky method [20,21] was used to prepare the glass-coated microwires with a diameter of 50–150 μm. After removing the glass sheath, the microwires were tested without any post heat treatments. The cross-section and surface morphologies of the microwire were investigated using a scanning electron microscope (SEM, Carl Zeiss, Oberkochen, Germany). The crystallographic orientation was studied using electron backscatter diffraction (EBSD, Carl Zeiss, Oberkochen, Germany), which was conducted at room temperature in the SEM. Synchrotron high-energy X-ray diffraction (HEXRD) experiments were conducted using a monochromatic X-ray beam with a wavelength of 0.1173 Å at the 11-ID-C beam line of the Advanced Photon Source at the Argonne National Laboratory, USA.

Mechanical tests were conducted in tension using a dynamic mechanical analyzer (DMA, TA Instruments, New Castle, DE, USA) with a maximum load of 18 N equipped with a closed furnace. The stress–strain curves were measured by force control at a loading–unloading rate of 50 MPa/min. The force was measured by using a load cell with a high resolution (10^{-5} N). The strain was determined by cross-head displacement using a high-resolution linear optical encoder that has a displacement resolution of 1 nm. The total length and gauge length of the samples for mechanical tests were about 12.2 mm and 7.0 mm, respectively. Magnetization as a function of temperature ($M(T)$) under constant magnetic fields and magnetization as a function of magnetic field ($M(H)$) at constant temperatures were measured by a physical property measurement system (PPMS, Quantum Design, San Diego, CA, USA). The $M(H)$ curves were measured by the standard loop process at different constant temperatures during two consecutive cycles of increasing the field to 5 T and then decreasing the field to 0 T. Before measuring the $M(H)$ curves at each temperature, the microwire was cooled to 135 K. This temperature was held for 1 min and then increased to the test temperature. The mass of the samples for the $M(T)$ and $M(H)$ measurements was 1.41 mg.

3. Results

3.1. Microstructure and Crystal Structure

Figure 1a shows the cross-section morphology of the $Ni_{50}Mn_{34}Fe_3In_{13}$ microwire, which indicates that the microwire had a regular, circular cross-section. The surface morphology of the $Ni_{50}Mn_{34}Fe_3In_{13}$ microwire is displayed in Figure 1b, which demonstrates that the microwire had a smooth surface and uniform diameter. The EBSD orientation map of the $Ni_{50}Mn_{34}Fe_3In_{13}$ wire was measured at room temperature and is shown in Figure 1c. No obvious grain boundaries can be observed in the sample used for EBSD measurement. This implies that the constraints of grain boundaries have been much reduced in the microwire—the grain size was as large as several millimeters [23].

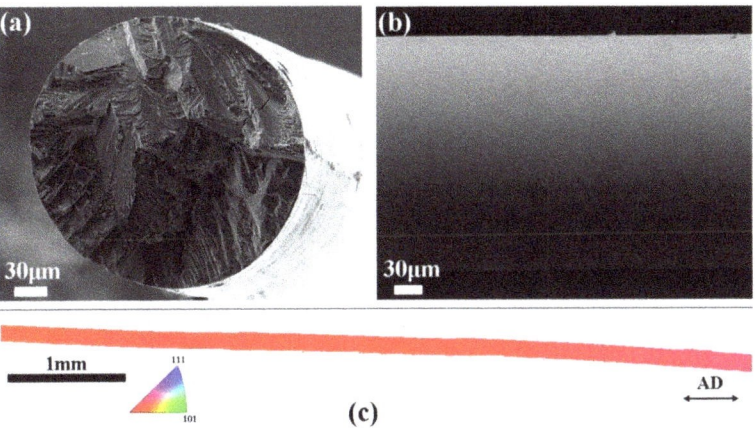

Figure 1. Microstructure of the $Ni_{50}Mn_{34}Fe_3In_{13}$ microwire at room temperature. (**a**,**b**) SEM images of the cross-section (**a**) surface (**b**); (**c**) Electron backscatter diffraction (EBSD) orientation map presented in inverse pole figure mode; the legend of a stereographic triangle (parallel to the wire axis direction AD) is also shown.

Figure 2a,b show the HEXRD patterns recorded at 298 K and 110 K, respectively, during the cooling of the $Ni_{50}Mn_{34}Fe_3In_{13}$ microwire. The crystal structures of austenite and martensite were identified with the help of the software PowderCell [25]. The pattern collected at 298 K in Figure 2a can be well indexed according to the cubic $L2_1$ Heusler structure (space group $Fm\bar{3}m$, No. 225) with lattice parameter a_A = 5.970 Å. As can be seen, besides the strong diffraction peaks of (220), (422) and (400), the superlattice reflections of (111), (311) and (331) can also be observed (see the inset of Figure 2a), which are characteristic of the $L2_1$ structure. At 110 K, the main diffraction peaks in Figure 2b can be well indexed according to the six-layered modulated (6M) structure of martensite (space group $P2/m$, No. 10) with lattice parameters a_{6M} = 4.395 Å, b_{6M} = 5.622 Å, c_{6M} = 25.824 Å and β = 92.10°. However, several other small peaks can also be seen in Figure 2b. These small peaks can be well indexed according to the cubic L2$_1$ structure of austenite with lattice parameter a_A = 5.955 Å. This implies that a tiny amount of austenite was retained at this temperature.

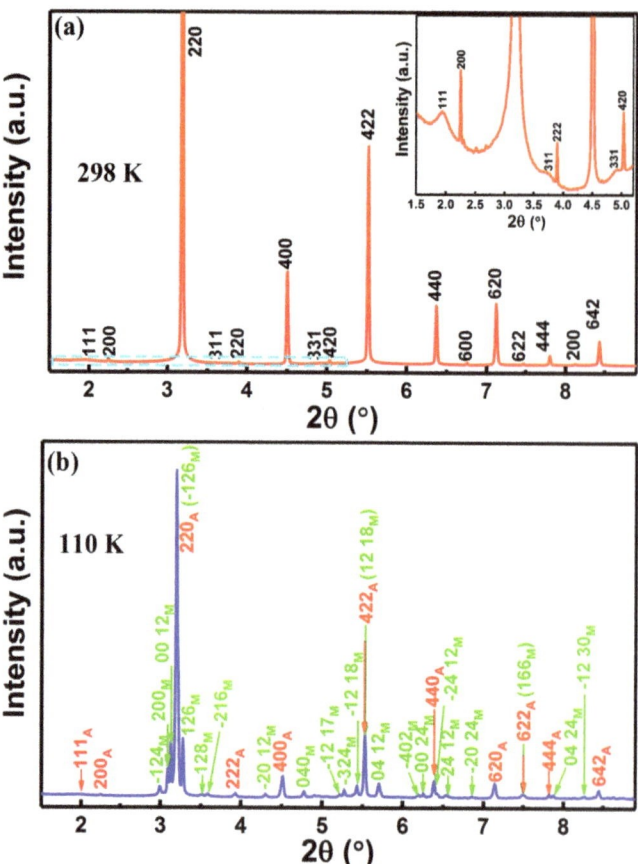

Figure 2. High-energy X-ray diffraction (HEXRD) patterns recorded at the temperatures of (a) 298 K and (b) 110 K for the Ni$_{50}$Mn$_{34}$Fe$_3$In$_{13}$ microwire. The inset in (a) displays the magnified view of the pattern in the 2θ range of 1.5–5.5°. "A" and "M" in the indices in (b) denote austenite and martensite, respectively.

Based on the geometric nonlinear theory of martensite, it is possible to evaluate the geometric compatibility between martensite and austenite. The middle eigenvalue λ_2 of the transformation stretch matrix **U**, which is used to characterize the geometric compatibility, can be computed with the algorithms reported in [26,27]. With the lattice parameters of austenite and martensite determined above, the λ_2 for the Ni$_{50}$Mn$_{34}$Fe$_3$In$_{13}$ microwire was determined to be 1.0083, which is close to 1. This implies a good geometric compatibility between martensite and austenite, which could explain the small thermal hysteresis of 8.9 K in the present microwire. This is because good geometric compatibility usually leads to a small thermal hysteresis [26].

3.2. Stress-Induced Phase Transformation and Superelasticity

In order to study the superelasticity that results from stress-induced martensitic transformation in the microwire, stress–strain curves at different constant temperatures were recorded. Figure 3a displays the tensile stress–strain curves at different constant temperatures in the temperature range of 213–283 K for the $Ni_{50}Mn_{34}Fe_3In_{13}$ microwire. The determination of the critical stress for stress-induced martensitic transformation σ_{cr}, stress hysteresis of the stress-induced martensitic transformation $\Delta\sigma$, irrecoverable strain after unloading ε_{irr}, superelastic strain ε_{se} and elastic strain ε_{el}, is illustrated in Figure 3b. In Figure 3a, at 213 K, two stress plateaus can be observed during loading, which indicated that inter-martensitic transformation occurred at this temperature. An irrecoverable strain ε_{irr} of 1.3% was observed after unloading when the maximum applied strain was 21.1%. This may be because a small amount of stress-induced martensite was stabilized and failed to transform back to austenite after unloading, as the test temperature (213 K) was slightly higher than the austenite transformation finish temperature A_f. This conjecture was confirmed by the recovery of the strain (1.3%) after heating to a higher temperature of 233 K. Encouragingly, the microwire exhibits excellent tensile superelasticity with almost no residual strain after unloading at higher temperatures (in the range between 233 and 283 K). Strikingly, a giant recoverable strain ε_{rec} of 20.3% was achieved in this temperature range, which is the highest recoverable strain reported so far in Ni-Mn-based SMAs. This value is much higher than that of Ni-Ti wire (approximately 11.5%), which is used for practical applications at present [28]. The temperature dependence of the critical stress for stress-induced martensitic transformation σ_{cr} is shown in Figure 3c. It can be seen that the critical stress increased linearly with the increase of temperature at a rate of 1.69 MPa/K.

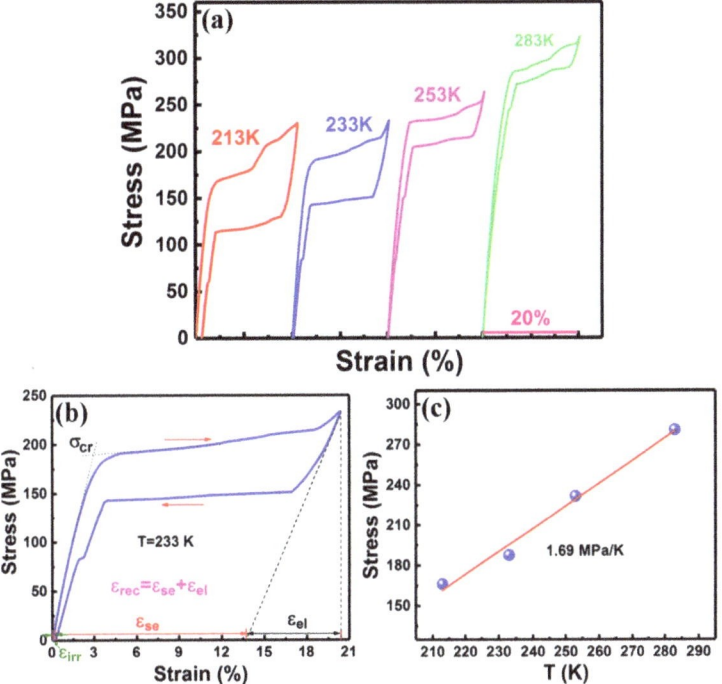

Figure 3. (a) Tensile stress–strain curves recorded at different constant temperatures in the range of 213–283 K for the $Ni_{50}Mn_{34}Fe_3In_{13}$ microwire. (b) Tensile stress–strain curve measured at 233 K with the determination of the following parameters: the critical stress for stress-induced martensitic transformation σ_{cr}, the stress hysteresis of the stress-induced martensitic transformation $\Delta\sigma$, the irrecoverable strain after unloading ε_{irr}, the superelastic strain ε_{se} and the elastic strain ε_{el}. The total recoverable strain ε_{rec} is the sum of ε_{se} and ε_{el}. (c) Temperature dependence of the critical stress for stress-induced martensitic transformation σ_{cr}.

The shape memory effect was examined by load-biased thermal cycling tests under different constant stresses. The strain–temperature curves recorded under stress levels of 25, 75, 125 and 225 MPa for the $Ni_{50}Mn_{34}Fe_3In_{13}$ microwire are shown in Figure 4. The recoverable strain ε_{rec} and irrecoverable strain ε_{irr} were determined and are illustrated in the figure. As can be seen, the strain associated with martensitic transformation was completely recovered after the cooling–heating cycle when the applied stresses were not higher than 75 MPa. The recoverable strains were high; they were 5.0% under 25 MPa and 8.9% under 75 MPa. The irrecoverable strain ε_{irr} occurred when the stress increased to above 125 MPa. In spite of this, ε_{rec} continued to increase as the stress increased, and amounted to 9.5% under 125 MPa. Strikingly, when the stress increased to 225 MPa, a ε_{rec} as high as 13.9% was achieved. This is the highest shape memory strain that has been reported to date in Ni-Mn-based SMAs, which is of great importance to realizing a large stroke in actuator applications [2,29].

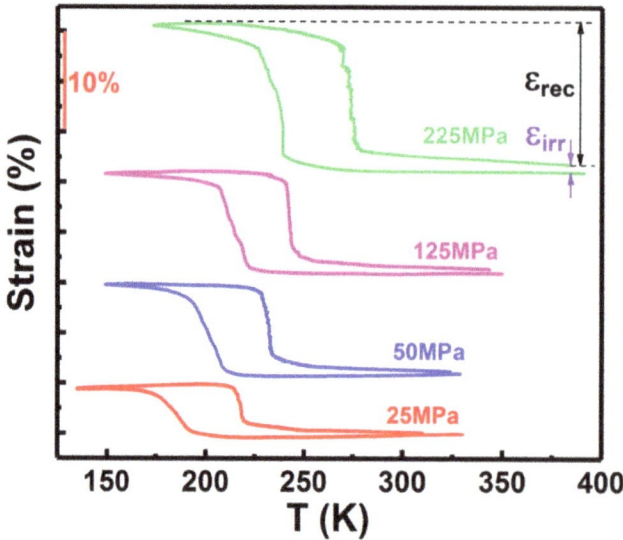

Figure 4. Strain–temperature curves measured under different constant stresses for the $Ni_{50}Mn_{34}Fe_3In_{13}$ microwire.

3.3. Magnetic-Field-Induced Phase Transformation and Magnetocaloric Effect

In order to study the magnetic properties of the $Ni_{50}Mn_{34}Fe_3In_{13}$ microwire, the $M(T)$ curves were measured. The $M(T)$ curves measured under 0.05 T and 5 T are shown in Figure 5a. As can be seen, the high-temperature austenitic phase and low-temperature martensitic phase are ferromagnetic and weak magnetic, respectively. During cooling, a major part of the austenite transformed into martensite when the temperature range was: (1) between 184 K and 182 K under 0.05 T; and (2) between 164 K and 162 K under 5 T, respectively. Upon further cooling, the remaining austenite continuously and gradually transformed into martensite. The dM/dT as a function of temperature, which is derived from the $M(T)$ curves in Figure 5a, is illustrated in Figure 5b. The martensitic transformation temperature (T_M) and the reverse transformation temperature (T_A) were determined by the temperatures corresponding to the maximum dM/dT values during cooling and heating, respectively. The T_M under the magnetic fields of 0.05 T and 5 T were 182.7 K and 162.2 K, respectively, while the T_A under 0.05 T and 5 T were 197.8 K and 178.3 K, respectively. As can be seen from Figure 5b, all the phase transformation temperatures decreased under 5 T when compared to those under 0.05 T. This may be due to the stabilization of austenite with a higher magnetization by the applied magnetic field. Specifically, T_A decreased by 19.5 K when the applied field was 5 T. Thus, the magnetic field dependence of T_A, $\Delta T_A/\mu_0\Delta H$, was −3.94 K/T.

The change in transformation temperature (ΔT_A) induced by the change in magnetic field ($\mu_0\Delta H$) usually follows the Clausius–Clapeyron relation [12,30,31]:

$$\frac{\Delta T_A}{\mu_0\Delta H} = -\frac{\Delta M}{\Delta S_{tr}} \qquad (1)$$

The transformation entropy change ΔS_{tr} estimated from the endothermic peak of the reverse transformation from the differential scanning calorimetry curve (not shown here) is 18.1 J kg^{-1} K^{-1}. The magnetization difference ΔM obtained from the $M(T)$ curve under 5 T was 68.0 A m^2 kg^{-1}. Therefore, $\Delta M/\Delta S_{tr}$ is 3.76 K/T, which is consistent with the experimental value of $\Delta T_A/\mu_0\Delta H$ (3.94 K/T). The obvious decrease in phase transformation temperatures indicated that a metamagnetic first-order phase transformation from martensite to austenite could be induced if a magnetic field was applied at a temperature that was close to the reverse transformation temperature.

To verify if the magnetic-field-induced phase transformation could occur in the Ni$_{50}$Mn$_{34}$Fe$_3$In$_{13}$ microwire and to examine the reversibility of the transformation, the $M(H)$ curves at different temperatures were measured. The $M(H)$ curves measured during two cycles of ascending and descending magnetic fields at different temperatures in the range 166–186 K are shown in Figure 5c. The thin lines and thick lines represent the first cycle and the second cycle of the ascending and descending magnetic fields, respectively. A rapid increase in magnetization was observed in the initial low-field range (below 0.2 T) at all the test temperatures. This may be due to the initial coexistence of weak magnetic martensite and a small amount of ferromagnetic austenite before the magnetic field was applied. As the magnetic field further increased, a jump in magnetization was observed at the critical field $\mu_0 H_{cr}$, particularly in the temperature range 178–186 K. This phenomenon implies that a strong metamagnetic first-order phase transformation from weak magnetic martensite to ferromagnetic austenite can be induced by the magnetic field. It is worth mentioning that the critical field $\mu_0 H_{cr}$ decreased as the temperature increased. Figure 5c indicates that only a portion of the phase transformation was induced under the magnetic field of 5 T at 166 K and 175 K. However, in the temperature range 178–186 K, the saturation of magnetization was observed at high fields. This indicates that the sample could transform completely into austenite at 5 T.

When comparing the first and second cycles of $M(H)$ curves, in the temperature range from 178 to 186 K, during increasing the magnetic field, the magnetization in the low-field region of the second cycle was slightly higher than that in the first cycle. This indicated that a small amount of austenite induced in first field cycle was retained and did not transform back into martensite during the decreasing magnetic field in the first cycle. The curve recorded during the decreasing field in the first cycle almost overlapped with that recorded during the decreasing field in the second cycle. In addition, the $M(H)$ curve recorded in the low magnetic field range during the increasing field in the second cycle was consistent with that during the decreasing field in the second cycle. These phenomena indicated that the part of retained austenite (only a tiny portion) after descending field in the first cycle did not participate in the subsequent transformation any more, and thus a reversible transformation between the martensite transformed back in the first cycle and the austenite could happen in the second and subsequent cycles. This is similar to the cases in [23,32]. As a result, a reversible first-order phase transformation, induced by a magnetic field, was achieved in the Ni$_{50}$Mn$_{34}$Fe$_3$In$_{13}$ microwire.

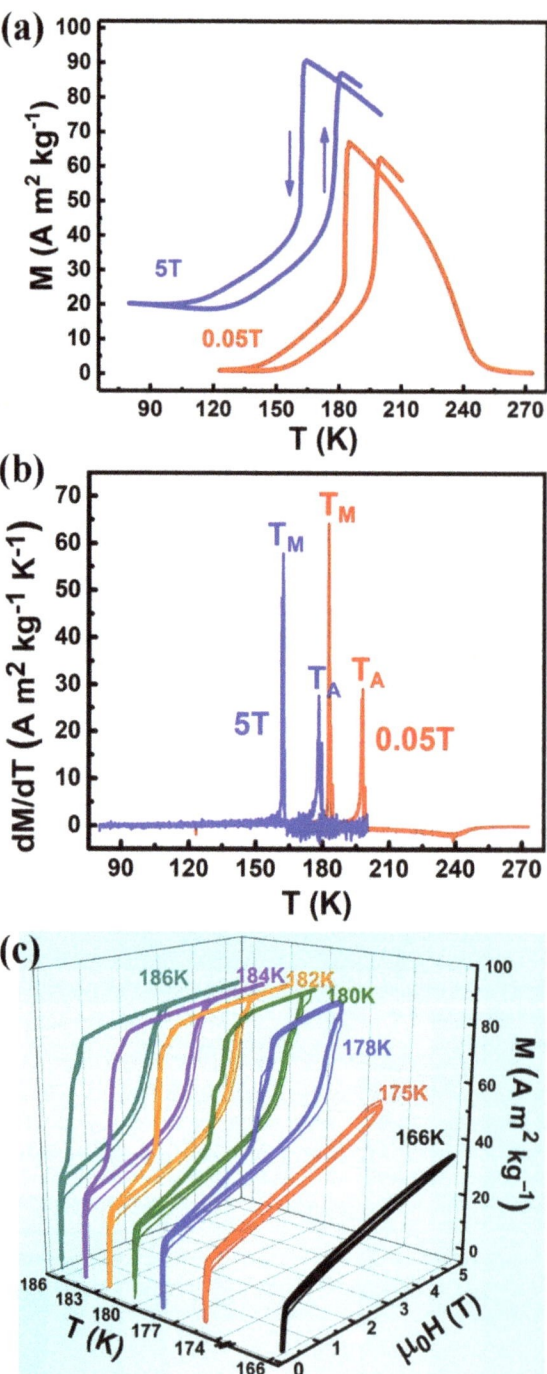

Figure 5. (a) $M(T)$ curves measured under 0.05 T and 5 T for the $Ni_{50}Mn_{34}Fe_3In_{13}$ microwire, (b) dM/dT derived from the curves in (a) shown as a function of temperature and (c) $M(H)$ curves measured during the first (thin lines) and second (thick lines) cycles of ascending and descending magnetic fields at different constant temperatures for the $Ni_{50}Mn_{34}Fe_3In_{13}$ microwire.

Magnetically driven multifunctional properties, such as magnetic superelasticity, magnetothermal conductivity, magnetoresistance and magnetocaloric effect, were anticipated in this $Ni_{50}Mn_{34}Fe_3In_{13}$ microwire. This is based on the reversible first-order phase transformation that was induced by a magnetic field. The magnetocaloric effect was estimated as an example of the multifunctional properties mentioned above. The reversible magnetic-field-induced entropy change (ΔS_m) was estimated from the $M(H)$ curves in Figure 5c. In the second cycle, the magnetic-field-induced phase transformation was reversible and, thus, the resultant ΔS_m was reversible as well. Therefore, the $M(H)$ curves in the second cycle were used to compute the reversible ΔS_m.

The critical field $\mu_0 H_{cr}$ for magnetic-field-induced phase transformation, extracted from the $M(H)$ curves in the second cycle, is shown as a function of temperature in Figure 6a. Linear fitting of the $\mu_0 H_{cr}$ vs. T data yielded a slope of -0.237 T/K, which was in accordance with the experimental $\mu_0 \Delta H / \Delta T_A$ value (-0.254 T/K). As a tiny amount of austenite coexists with martensite before applying the magnetic field, the estimation of ΔS_m using the Maxwell relation may lead to spurious results [33]. In contrast, estimation using the Clausius–Clapeyron relation could yield the correct value of ΔS_m even in the case of the coexistence of two phases, since ΔS_m in the Clausius–Clapeyron relation is directly connected to the field-induced magnetization difference at any given temperature. Therefore, it is more suitable to use the Clausius–Clapeyron relation to estimate the reversible ΔS_m. The ΔS_m can be estimated by the following relation [34–36]:

$$\Delta S_m = -\Delta M' \frac{d(\mu_0 H_{cr})}{dT} \qquad (2)$$

in which $\Delta M'$ is the magnetization difference between the magnetization values at the final and initial fields. Since the magnetization rapidly changes in the $M(H)$ curves below 0.2 T, which could lead to numerical instabilities, the initial magnetic field selected was 0.2 T. The $d(\mu_0 H_{cr})/dT$ is -0.237 T/K, as mentioned above.

The reversible field-induced entropy change ΔS_m for a magnetic field change from 0.2 T to 5 T at different temperatures is displayed in Figure 6b. It can be seen that an inverse magnetocaloric effect was achieved as the ΔS_m values at all the temperatures are positive. The maximum reversible ΔS_m for a magnetic field change from 0.2 T to 5 T is as high as 15.1 J kg^{-1} K^{-1}, which is higher than that in typical MMSMA microwires [23,33]. The large reversible magnetocaloric effect and high specific surface area confer on the present microwire high potential for magnetic refrigeration applications.

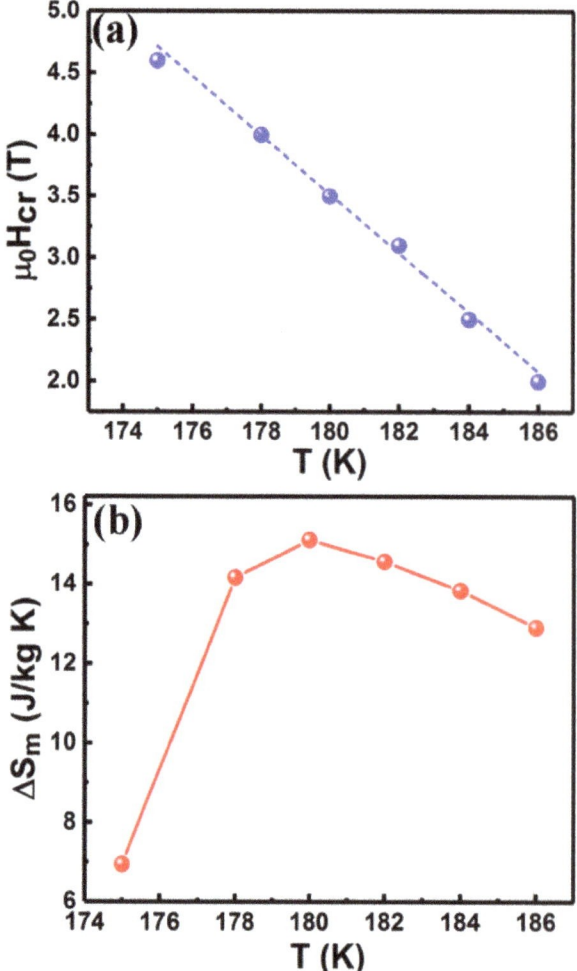

Figure 6. (a) Critical magnetic field for magnetic-field-induced phase transformation ($\mu_0 H_{cr}$) that was extracted from the $M(H)$ curves in the second cycle in Figure 5b, shown as a function of temperature for the $Ni_{50}Mn_{34}Fe_3In_{13}$ microwire. The dashed line represents the linear fitting of the data (denoted by symbols). (b) Reversible magnetic-field-induced entropy change ΔS_m as a function of temperature for the field change from 0.2 T to 5 T for the $Ni_{50}Mn_{34}Fe_3In_{13}$ microwire.

4. Discussion

The magnetostress is an important property for magnetic actuation applications and it is usually defined as the change in the critical stress for stress-induced martensitic transformation under a given magnetic field [37]. In Ni-Mn-based MMSMAs, since the austenite is ferromagnetic and the martensite is weak magnetic, the magnetic field favors austenite. This leads to austenite stabilization and an increase in critical stress for stress-induced martensitic transformation at a given temperature; the magnetic field and the stress act in opposite directions under simultaneously applied magnetic field and stress.

The Zeeman energy, which arises from the difference between the saturation magnetizations of austenite and martensite, is the driving force for magnetic-field-induced first-order phase transformation [38]. Zeeman energy continuously increases with an increasing magnetic field. Therefore, it is possible to achieve a large output magnetostress in the present Ni$_{50}$Mn$_{34}$Fe$_3$In$_{13}$ microwire, considering the large magnetization difference between austenite and martensite.

It is possible to predict the magnetostress as a function of applied field if the change in critical stress for stress-induced phase transformation with temperature ($\Delta\sigma/\Delta T$) and the change in transformation temperature with applied field ($\Delta T/\mu_0\Delta H$) are known. The change in stress with applied magnetic field ($\Delta\sigma/\mu_0\Delta H$) can be approximated as follows [37,38]:

$$\frac{\Delta\sigma}{\mu_0\Delta H} = -\frac{\Delta\sigma}{\Delta T} \times \frac{\Delta T}{\mu_0\Delta H}. \tag{3}$$

The $\Delta\sigma/\Delta T$ and $\Delta T_A/\mu_0\Delta H$ for the present Ni$_{50}$Mn$_{34}$Fe$_3$In$_{13}$ microwire are 1.69 MPa/K and -3.94 K/T, respectively, as determined before. With Equation (3), $\Delta\sigma/\mu_0\Delta H$ was estimated to be 6.66 MPa/T for the Ni$_{50}$Mn$_{34}$Fe$_3$In$_{13}$ microwire. Therefore, a large magnetic work output is expected from the present microwire, showing its potential for magnetic actuation applications. Under a magnetic field of 1 T, it is possible to obtain a magnetostress of 6.66 MPa and under 5 T, the magnetostress would be 33.3 MPa.

The present Ni$_{50}$Mn$_{34}$Fe$_3$In$_{13}$ microwire exhibited both magnetic-field-induced phase transformation and stress-induced martensitic transformation. Since stress-induced martensitic transformation occurred between 213 K and 283 K, a considerable elastocaloric effect was anticipated in this temperature range. Based on the magnetic-field-induced transformation, a large reversible magnetocaloric effect was achieved and other magnetoresponsive properties such as magnetic-field-induced strain and magnetoresistance were expected. Owing to the coupling between elastic deformation and magnetization, the elastomagnetic effect [39–41] could also be achieved in the microwire. Since the strain change could be detected by monitoring the change in magnetization, the microwire could also be used in non-contact strain sensors.

5. Conclusions

A Ni-Fe-Mn-In microwire exhibiting giant tensile superelasticity and magnetic-field-induced first-order phase transformation was developed. This Ni$_{50}$Mn$_{34}$Fe$_3$In$_{13}$ microwire shows a giant recoverable strain of more than 20% as a result of stress-induced martensitic transformation in the temperature range of 233–283 K. This represents the highest recoverable strain reported heretofore in Ni-Mn-based SMAs. In addition, the present microwire exhibits a large shape memory effect with a recoverable strain of up to 13.9% under a constant tensile stress of 225 MPa. These properties contrast with those in the bulk MMSMAs that barely show any tensile deformability. Due to the considerable magnetization difference between martensite and austenite, magnetic-field-induced first-order phase transformation is realized in this microwire. Thus, a series of magneto-responsive properties were anticipated. Indeed, a large reversible magnetocaloric effect, with an isothermal entropy change ΔS_m of 15.1 J kg^{-1} K^{-1} for a field change from 0.2 T to 5 T, was obtained from this microwire. The achievement of magnetic-field and stress-induced transformations confers on this Ni$_{50}$Mn$_{34}$Fe$_3$In$_{13}$ microwire great potential for application in miniature multifunctional devices.

Author Contributions: Conceptualization, D.C. and Z.C.; methodology, Z.C., S.L. (Shilei Li) and Y.Z.; validation, Z.C., D.C., S.L. (Shilei Li), Y.W. and Y.R.; formal analysis, Z.C., Y.Z., Y.C., S.L. (Shaohui Li) and C.S.; investigation, Z.C., Y.Z., and S.L. (Shengwei Li); writing—original draft preparation, Z.C.; writing—review and editing, D.C. and Z.C. All authors have read and agreed to the published version of the manuscript.

Funding: Financial support from the National Natural Science Foundation of China (Nos. 51731005, 51822102, 52001005, 51831003 and 51527801) and the Fundamental Research Funds for the Central Universities (No. FRF-TP-18-008C1) is gratefully acknowledged. This work is also supported by the Funds for Creative Research Groups of China (No. 51921001). Use of the Advanced Photon Source was supported by the U.S. Department of Energy, Office of Science, Office of Basic Energy Science, under Contract No. DE-AC02-06CH11357.

Data Availability Statement: The data that support the findings of this study are available from the corresponding author, upon reasonable request.

Conflicts of Interest: The authors declare no conflict of interest.

References

1. Otsuka, K.; Wayman, C.M. *Shape Memory Materials*; Cambridge University Press: Cambridge, UK, 1998.
2. Jani, J.M.; Leary, M.; Subic, A.; Gibson, M.A. A review of shape memory alloy research, applications and opportunities. *Mater. Des.* **2014**, *56*, 1078–1113. [CrossRef]
3. Sun, L.; Huang, W.M.; Ding, Z.; Zhao, Y.; Wang, C.C.; Purnawali, H.; Tang, C. Stimulus-responsive shape memory materials: A review. *Mater. Des.* **2012**, *33*, 577–640. [CrossRef]
4. Manjaiah, M.; Narendranath, S.; Basavarajappa, S. Review on non-conventional machining of shape memory alloys. *Trans. Nonferrous Met. Soc. China* **2014**, *24*, 12–21. [CrossRef]
5. Wu, M.H.; Schetky, L. Industrial Applications for Shape Memory Alloys. In Proceedings of the International Conference on Shape Memory and Superelastic Technologies, Pacific Grove, CA, USA, 30 April–4 May 2000; pp. 171–182.
6. Stoeckel, D. Shape memory actuators for automotive applications. *Mater. Des.* **1990**, *11*, 302–307. [CrossRef]
7. Velmurugan, C.; Senthilkumar, V.; Dinesh, S.; Arulkirubakaran, D. Review on phase transformation behavior of NiTi shape memory alloys. *Mater. Today Proc.* **2018**, *5*, 14597–14606. [CrossRef]
8. Choudhary, N.; Kaur, D. Shape memory alloy thin films and heterostructures for MEMS applications: A review. *Sens. Actuators A* **2016**, *242*, 162–181. [CrossRef]
9. Petrini, L.; Migliavacca, F. Biomedical applications of shape memory alloys. *J. Met.* **2011**, *2011*, 501483. [CrossRef]
10. Krenke, T.; Duman, E.; Acet, M.; Wassermann, E.F.; Moya, X.; Mañosa, L.; Planes, A.; Suard, E.; Ouladdiaf, B. Magnetic superelasticity and inverse magnetocaloric effect in Ni-Mn-In. *Phys. Rev. B* **2007**, *75*, 104414. [CrossRef]
11. Mañosa, L.; Moya, X.; Planes, A.; Aksoy, S.; Acet, M.; Wassermann, E.; Krenke, T. Magnetostrain in Multifunctional Ni-Mn Based Magnetic Shape Memory Alloys. *Mater. Sci. Forum* **2008**, *583*, 111–117. [CrossRef]
12. Kainuma, R.; Imano, Y.; Ito, W.; Sutou, Y.; Morito, H.; Okamoto, S.; Kitakami, O.; Oikawa, K.; Fujita, A.; Kanomata, T.; et al. Magnetic-field-induced shape recovery by reverse phase transformation. *Nature* **2006**, *439*, 957–960. [CrossRef]
13. Pathak, A.K.; Dubenko, I.; Pueblo, C.; Stadler, S.; Ali, N. Magnetoresistance and magnetocaloric effect at a structural phase transition from a paramagnetic martensitic state to a paramagnetic austenitic state in $Ni_{50}Mn_{36.5}In_{13.5}$ Heusler alloys. *Appl. Phys. Lett.* **2010**, *96*, 172503. [CrossRef]
14. Huang, L.; Cong, D.Y.; Ma, L.; Nie, Z.H.; Wang, Z.L.; Suo, H.L.; Ren, Y.; Wang, Y.D. Large reversible magnetocaloric effect in a Ni-Co-Mn-In magnetic shape memory alloy. *Appl. Phys. Lett.* **2016**, *108*, 032405. [CrossRef]
15. Liu, J.; Woodcock, T.G.; Scheerbaum, N.; Gutfleisch, O. Influence of annealing on magnetic field-induced structural transformation and magnetocaloric effect in Ni-Mn-In-Co ribbons. *Acta Mater.* **2009**, *57*, 4911–4920. [CrossRef]
16. Liu, J.; Gottschall, T.; Skokov, K.P.; Moore, J.D.; Gutfleisch, O. Giant magnetocaloric effect driven by structural transitions. *Nat. Mater.* **2012**, *11*, 620–626. [CrossRef]
17. Ueland, S.M.; Chen, Y.; Schuh, C.A. Oligocrystalline Shape Memory Alloys. *Adv. Funct. Mater.* **2012**, *22*, 2094–2099. [CrossRef]
18. Dunand, D.C.; Müllner, P. Size effects on magnetic actuation in Ni-Mn-Ga shape-memory alloys. *Adv. Mater.* **2011**, *23*, 216–232. [CrossRef]
19. Juan, J.S.; Nó, M.L.; Schuh, C.A. Superelasticity and Shape Memory in Micro- and Nanometer-scale Pillars. *Adv. Mater.* **2008**, *20*, 272–278. [CrossRef]
20. Chiriac, H.; Óvári, T.A. Amorphous glass-covered magnetic wires: Preparation, properties, applications. *Prog. Mater. Sci.* **1996**, *40*, 333–407. [CrossRef]
21. Vázquez, M.; Chiriac, H.; Zhukov, A.; Panina, L.; Uchiyama, T. On the state-of-the-art in magnetic microwires and expected trends for scientific and technological studies. *Phys. Status Solidi A* **2011**, *208*, 493–501. [CrossRef]
22. Qian, M.F.; Zhang, X.X.; Witherspoon, C.; Sun, J.F.; Müllner, P. Superelasticity and shape memory effects in polycrystalline Ni-Mn-Ga microwires. *J. Alloys Compd.* **2013**, *577*, S296–S299. [CrossRef]
23. Chen, Z.; Cong, D.; Sun, X.; Zhang, Y.; Yan, H.; Li, S.; Li, R.; Nie, Z.; Ren, Y.; Wang, Y. Magnetic field-induced magnetostructural transition and huge tensile superelasticity in an oligocrystalline Ni–Cu–Co–Mn–In microwire. *IUCrJ* **2019**, *6*, 843–853. [CrossRef]
24. Li, F.Q.; Qu, Y.H.; Yan, H.L.; Chen, Z.; Cong, D.Y.; Sun, X.M.; Li, S.H.; Wang, Y.D. Giant tensile superelasticity originating from two-step phase transformation in a Ni-Mn-Sn-Fe magnetic microwire. *Appl. Phys. Lett.* **2018**, *113*, 112402. [CrossRef]
25. Kraus, W.; Nolze, G. POWDER CELL–a program for the representation and manipulation of crystal structures and calculation of the resulting X-ray powder patterns. *J. Appl. Crystallogr.* **1996**, *29*, 301–303. [CrossRef]
26. Song, Y.T.; Chen, X.; Dabade, V.; Shield, T.W.; James, R.D. Enhanced reversibility and unusual microstructure of a phase-transforming material. *Nature* **2013**, *502*, 85–88. [CrossRef]
27. Hane, K.F.; Shield, T.W. Microstructure in the cubic to monoclinic transition in titanium-nickel shape memory alloys. *Acta Mater.* **1999**, *47*, 2603–2617. [CrossRef]
28. Liang, X.; Xiao, F.; Jin, M.; Jin, X.; Fukuda, T.; Kakeshita, T. Elastocaloric effect induced by the rubber-like behavior of nanocrystalline wires of a Ti-50.8Ni (at.%) alloy. *Scr. Mater.* **2017**, *134*, 42–46. [CrossRef]
29. Ma, J.; Karaman, I.; Noebe, R.D. High temperature shape memory alloys. *Int. Mater. Rev.* **2011**, *55*, 257–315. [CrossRef]

30. Kustov, S.; Corró, M.L.; Pons, J.; Cesari, E. Entropy change and effect of magnetic field on martensitic transformation in a metamagnetic Ni-Co-Mn-In shape memory alloy. *Appl. Phys. Lett.* **2009**, *94*, 191901. [CrossRef]
31. Cong, D.Y.; Roth, S.; Schultz, L. Magnetic properties and structural transformations in Ni-Co-Mn-Sn multifunctional alloys. *Acta Mater.* **2012**, *60*, 5335–5351. [CrossRef]
32. Qu, Y.H.; Cong, D.Y.; Chen, Z.; Gui, W.Y.; Sun, X.M.; Li, S.H.; Ma, L.; Wang, Y.D. Large and reversible inverse magnetocaloric effect in $Ni_{48.1}Co_{2.9}Mn_{35.0}In_{14.0}$ metamagnetic shape memory microwire. *Appl. Phys. Lett.* **2017**, *111*, 192412. [CrossRef]
33. Liu, G.J.; Sun, J.R.; Shen, J.; Gao, B.; Zhang, H.W.; Hu, F.X.; Shen, B.G. Determination of the entropy changes in the compounds with a first-order magnetic transition. *Appl. Phys. Lett.* **2007**, *90*, 032507. [CrossRef]
34. Balli, M.; Fruchart, D.; Gignoux, D.; Zach, R. The "colossal" magnetocaloric effect in $Mn_{1-x}Fe_xAs$: What are we really measuring? *Appl. Phys. Lett.* **2009**, *95*, 072509. [CrossRef]
35. Qu, Y.H.; Cong, D.Y.; Sun, X.M.; Nie, Z.H.; Gui, W.Y.; Li, R.G.; Ren, Y.; Wang, Y.D. Giant and reversible room-temperature magnetocaloric effect in Ti-doped Ni-Co-Mn-Sn magnetic shape memory alloys. *Acta Mater.* **2017**, *134*, 236–248. [CrossRef]
36. Szymczak, R.; Nedelko, N.; Lewińska, S.; Zubov, E.; Sivachenko, A.; Gribanov, I.; Radelytskyi, I.; Dyakonov, K.; Ślawska-Waniewska, A.; Valkov, V.; et al. Comparison of magnetocaloric properties of the $Mn_{2-x}Fe_xP_{0.5}As_{0.5}$ (x = 1.0 and 0.7) compounds. *Solid. State Sci.* **2014**, *36*, 29–34. [CrossRef]
37. Karaca, H.E.; Karaman, I.; Basaran, B.; Ren, Y.; Chumlyakov, Y.I.; Maier, H.J. Magnetic Field-Induced Phase Transformation in NiMnCoIn Magnetic Shape-Memory Alloys-A New Actuation Mechanism with Large Work Output. *Adv. Funct. Mater.* **2009**, *19*, 983–998. [CrossRef]
38. Turabi, A.S.; Karaca, H.E.; Tobe, H.; Basaran, B.; Aydogdu, Y.; Chumlyakov, Y.I. Shape memory effect and superelasticity of NiMnCoIn metamagnetic shape memory alloys under high magnetic field. *Scr. Mater.* **2016**, *111*, 110–113. [CrossRef]
39. Ausanio, G.; Hison, C.; Iannotti, V.; Luponio, C.; Lanotte, L. Elastomagnetic effect in novel elastic magnets. *J. Magn. Magn. Mater.* **2004**, *272–276*, 2069–2071. [CrossRef]
40. Lanotte, L.; Ausanio, G.; Iannotti, V.; Pepe, G.; Carotenuto, G.; Netti, P.; Nicolais, L. Magnetic and magnetoelastic effects in a composite material of Ni microparticles in a silicone matrix. *Phys. Rev. B* **2001**, *63*, 811–820. [CrossRef]
41. Tanaka, Y.; Himuro, Y.; Kainuma, R.; Sutou, Y.; Omori, T.; Ishida, K. Ferrous Polycrystalline Shape-Memory Alloy Showing Huge Superelasticity. *Science* **2010**, *327*, 1488–1490. [CrossRef]

Article

Elastocaloric and Magnetocaloric Effects Linked to the Martensitic Transformation in Bulk $Ni_{55}Fe_{11}Mn_7Ga_{27}$ Alloys Produced by Arc Melting and Spark Plasma Sintering

J. D. Navarro-García [1,*], J. P. Camarillo-Garcia [2], F. Alvarado-Hernández [2], J. L. Sánchez Llamazares [1,*] and H. Flores-Zúñiga [1]

1. Instituto Potosino de Investigación Científica y Tecnológica A.C., Camino a la Presa San José 2055, Col. Lomas 4 Section, San Luis Potosí 78216, Mexico; horacio.flores@ipicyt.edu.mx
2. Unidad Académica de Ingeniería I, Universidad Autónoma de Zacatecas, Ramón López Velarde 801, Col. Centro, Zacatecas 98000, Mexico; jp.camarillo.garcia@uaz.edu.mx or jp.camarillo.garcia@gmail.com (J.P.C.-G.); ingenierofah@gmail.com (F.A.-H.)
* Correspondence: daniel.navarro@ipicyt.edu.mx (J.D.N.-G.); jose.sanchez@ipicyt.edu.mx (J.L.S.L.); Tel.: +52-444-8342000 (J.L.S.L.)

Abstract: The investigation of caloric effects linked to first-order structural transitions in Heusler-type alloys has become a subject of considerable current interest due to their potential utilization as refrigerants in solid-state cooling devices. This study is mainly motivated by the possibility of developing refrigeration devices of improved energy efficiency with a reduced environmental impact. We produced partially textured and isotropic bulk samples of the Heusler-type magnetic shape memory alloy $Ni_{55}Fe_{11}Mn_7Ga_{27}$ by arc melting and spark plasma sintering (SPS), respectively. Their structural, microstructural, and phase transition characteristics and magnetocaloric and elastocaloric effects, associated with first-order martensitic transformation (MT), were studied. The elemental chemical compositions of both samples were close to nominal, and a martensitic-like structural transformation appeared around room temperature with similar starting and finishing structural transition temperatures. At room temperature, austenite exhibited a highly ordered $L2_1$-type crystal structure. The partial grain orientation and isotropic nature of the arc-melted and SPS samples, respectively, were revealed by X-ray diffraction and SEM observations of the microstructure. For the arc-melted sample, austenite grains preferentially grew in the (100) direction parallel to the thermal gradient during solidification. The favorable effect of the texture on the elastocaloric response was demonstrated. Finally, due to its partial grain orientation, the arc-melted bulk sample showed superior values of maximum magnetic entropy change ($|\Delta S_M|^{max}$ = 18.6 Jkg^{-1}K^{-1} at 5 T) and elastocaloric adiabatic temperature change ($|\Delta T_{ad}^{me}|^{max}$ = 2.4 K at 120 MPa) to those measured for the SPS sample ($|\Delta S_M|^{max}$ = 8.5 Jkg^{-1}K^{-1} and ($|\Delta T_{ad}^{me}|^{max}$ = 0.8 K).

Keywords: Ni-Fe-Ga magnetic shape memory alloys; elastocaloric and magnetocaloric effects; martensitic transformation; spark plasma sintering

1. Introduction

The recent interest in the development of solid-state refrigeration devices as a more efficient alternative to conventional gas compression refrigeration has encouraged the investigation of the caloric effects linked to first-order phase transitions in different families of solids [1]. Among them, (Ni,Mn)-based Heusler-type alloys undergoing a diffusion-less martensitic transformation (MT) have been the subject of considerable interest due to their multifunctional nature [1]. Elastocaloric (eC) and magnetocaloric (MC) effects are thermal responses to external stimuli, mechanical or magnetic, respectively, characterized by the adiabatic temperature change ΔT_{ad} or the isothermal entropy change ΔS due to the application of a change in the stress $\Delta \sigma$ or the magnetic field $\mu_0 \Delta H$ [2]. Martensitic transformation (MT) in shape memory alloys (SMAs) may show a high sensitivity to the application of external uniaxial stress, which, in some cases, gives rise to a large eC effect, making them promising materials for the development of solid-state refrigeration devices [3–5]. However, depending on the alloy, a large load may be needed to induce the structural transformation compromising the potential applicability [6]. In magnetic shape memory alloys (MSMA), the structural change is accompanied by a change in the magnetic state of the martensite and austenite phases with the consequent magnetization change ΔM, which in turn may lead to a large MC effect [7]. Thus, the material is considered multicaloric if different caloric effects linked to the transition can be driven simultaneously, or sequentially, by the application of more than one external field or stimulus. Currently, there is a rising interest in the investigation of multicaloric materials [8].

(Ni, Mn)-based Heusler-type magnetic shape memory alloys have received particular attention due to their multicaloric capability [9]. The present work focuses on an MSMA derived from a $Ni_{55}Fe_{19}Ga_{26}$ alloy in which the Fe was partially substituted by Mn. For properly chosen compositions, MT in Ni-Fe-Ga alloys appears around room temperature (RT) [1,10]. The structural transition on cooling is from a cubic ferromagnetic austenite with a B2-type or the $L2_1$-type crystal structure, depending on the thermal treatment received above 973 K [11], to 10M or 14M modulated ferromagnetic martensite [10–13]. The two factors that motivated the present investigation were the good mechanical properties of this alloy system in comparison to other (Ni,Mn)-based alloys (which are further improved with the precipitation of the so-called γ phase [14]) and the evidence that the partial replacement of Fe by Mn increases the entropy of the MT without significantly modifying the MT temperatures or the γ low concentration present in the $Ni_{55}Fe_{19}Ga_{26}$ composition [15]. Moreover, Ni-Fe-Ga alloys can be easily obtained by a conventional melting technique, such as arc melting, followed by the thermal treatment required in order to produce the desired austenitic structure [16]. Another aspect addressed in this work was the consolidation of these alloys by spark plasma sintering (SPS). This pressure-assisted processing technique allows the fabrication of highly dense shaped pieces of a large variety of metallic and ceramic materials in a very short thermal processing time and at relatively low temperatures in comparison to those reported when conventional sintering is used [17,18]. The ability of the SPS technique to produce highly dense (Ni,Mn)-based alloys with improved mechanical resistance has been highlighted by several authors [18–21] example, in [22], this processing route was used to fabricate Ni-Co-Mn-Sn alloys with similar MT temperatures and magnetic properties to those obtained for the bulk alloy produced by induction melting [22]. To the best of our knowledge, (Ni-Fe-Ga)-based alloys have not yet been produced by SPS.

Although there have been several studies about the magnetic characterization of Ni-Fe-Ga alloys [23–27], the MC effect has rarely been reported, mainly because the maximum magnetic entropy change $|\Delta S_M|^{max}$ displayed by these alloys does not exceed ~5 J$kg^{-1}K^{-1}$ at $\mu_0 \Delta H$ = 5 T [15,28], with a single report of ~12 J$kg^{-1}K^{-1}$ [14]. However, eC effect reports on single crystals or textured alloys have recently been presented with large mechanically induced entropy change or adiabatic temperature change [3,14,29]. Therefore, in this investigation, we synthetized polycrystalline samples of the quaternary $Ni_{55}Fe_{11}Mn_7Ga_{27}$ alloy by arc melting and spark plasma sintering (SPS), and studied their microstructural, structural, magnetocaloric, and elastocaloric properties.

2. Materials and Methods

Arc-melted ingots of nominal composition $Ni_{55}Fe_{11}Mn_7Ga_{27}$ were synthetized from pure elements ($\geq$99.9%); a copper crucible with a flat base was used to promote vertical grain growth along the thermal gradient. Next, the samples were enclosed in quartz capsules in a UHP argon atmosphere to receive two subsequent thermal treatments: homogenization at 1273 K for 24 h and a chemical ordering treatment at 773 K for 1 h (both ending by quenching in iced water). From one of the arc melted samples, which will be referred to hereafter as "bulk", different specimens were cut with a low-speed diamond saw for the different studies performed. The other was manually pulverized with an agate mortar to be processed by SPS (i.e., SPS sample). The SPS process was carried out in a Labox-210 SPS system (from Sinter Land Inc., Niigata, Japan), using a 15 mm diameter graphite die, under vacuum at constant pressure of 30 MPa. The time evolution of the vertical displacement and temperature through the SPS process is shown in Figure 1. Notably, the vertical displacement continuously rises to 693 K (420 °C), due to the thermal expansion of the sample, whereas from this temperature and up to 1182 K (909 °C), a continuous shrinkage takes place for 9 min, denoting the occurrence of sintering. The obtained sintered disc showed a density ρ of 8.4×10^3 kg·m^{-3}; this was ~99% of the maximum expected value [30].

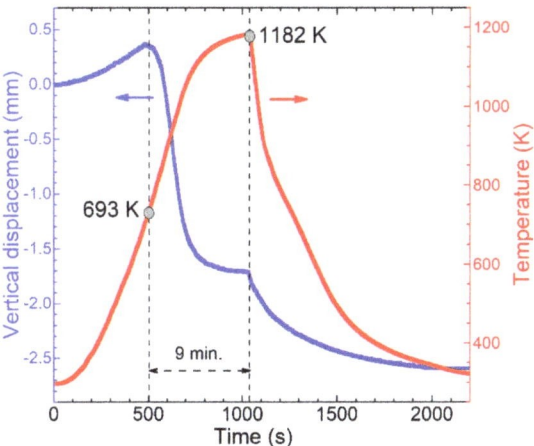

Figure 1. Vertical displacement and temperature versus time along the SPS process.

Differential scanning calorimetry (DSC) curves were measured on a TA-DSC Q200 at a T sweep rate of 10 Kmin^{-1} to determine the enthalpy ΔH_{st}, entropy ΔS_{st} and characteristic starting and finishing temperatures of the structural transformation. Martensitic and reverse martensitic starting and finishing temperatures are referred to as M_S and M_f, and A_S and A_f, respectively; the method for their determination is given below. The X-ray diffraction (XRD) analyses were performed in a Rigaku Smartlab high-resolution powder diffractometer (Cu-K$_\alpha$ radiation). The microstructural characterization was performed using a QUANTA FEG 250 scanning electron microscope (SEM) from FEI; the system was equipped with an EDS system from EDAX. Magnetization as a function of temperature $M(T)$ curves was measured at a temperature sweep rate of 1.0 Kmin^{-1} on parallelepiped-shaped specimens; for the arc-melted and SPS samples, the major axis of the parallelepipeds was parallel to the thermal gradient and pressing directions, respectively. The magnetic field was applied along the greatest length of the samples to minimize the internal demagnetizing field. The measurements were performed using the vibrating sample magnetometer option of a Quantum Design PPMS® Dynacool® system. Finally, in [2], the purpose-built equipment used for the direct measurement of the eC adiabatic temperature change ΔT_{ad}^{me} is described. This measurement was performed on parallelepiped-shaped specimens cut from the bulk and SPS samples; their physical dimensions were $4.3 \times 2.2 \times 2.1$ mm^3 and $2.2 \times 2.3 \times 2.1$ mm^3 (i.e., height, width, and length), respectively. As temperature probe, the system used a tiny K-type thermocouple (0.13 mm in diameter). Compressing uniaxial load change of 120 MPa was applied parallel to the thermal gradient during solidification and pressing direction for bulk and SPS samples, respectively.

3. Results and Discussion

Figure 2a shows the heating and cooling DSC curves for both samples. The exothermic and endothermic peaks reveal, respectively, the martensitic and austenitic transitions of both alloys. By determining the transformed fraction as a function of the temperature of one phase to the other for the structural transition in both directions (i.e., heating and cooling), the start and finish temperatures of the structural transformation were determined at 5% and 95% of the transformation percentage, respectively. Their values, together with the thermal hysteresis of the transformation ΔT_{hyst} (determined as $\Delta T_{\text{hyst}} = A_f - M_S$), are listed in Table 1. Notably, the structural transition in both samples appears around RT and their starting and finishing structural transition temperatures are similar. Table 1 also shows that, within the error of the determination, the averaged elemental chemical composition determined by EDS is similar in both samples and close to nominal. This result is relevant, since the transition temperatures in these (Ni-Fe-Ga)-based alloys are highly sensitive to composition alterations [31]. Consequently, both results highlight the effectiveness of the SPS technique at consolidating (Ni-Fe-Ga)-based alloys.

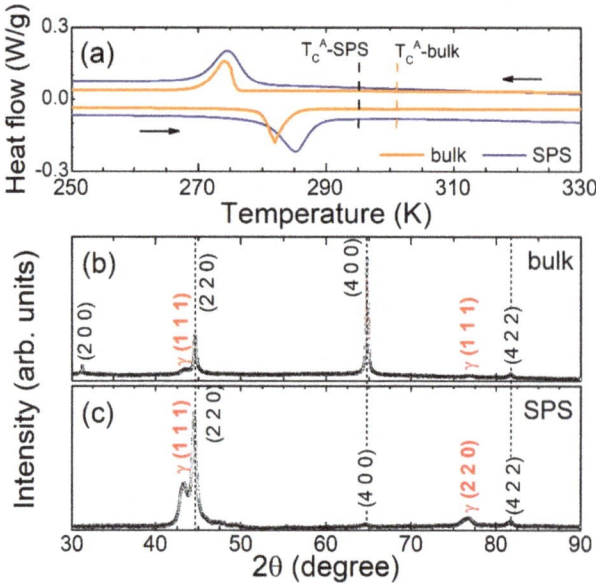

Figure 2. (a) DSC curves for bulk and SPS samples. Vertical dotted lines were drawn to indicate the Curie temperature of austenite for both samples (T_C^A-bulk and T_C^A-SPS, respectively). (b,c) Indexed XRD patterns for bulk and SPS samples, respectively.

Table 1. Characteristic starting and finishing temperatures, thermal hysteresis, structural enthalpy and entropy of the structural transformation, Curie temperature of austenite (determined from the minimum of the $dM/dT(T)$ curve measured at 5 mT), and experimental elemental chemical composition determined by EDS (at. %) for bulk arc-melted and SPS samples.

Sample	Method	M_S (K)	M_f (K)	A_S (K)	A_f (K)	ΔT_{hyst} (K)	ΔH_{st} (Jg^{-1})	ΔS_{st} (Jkg^{-1}K^{-1})	T_C^A (K)	Experimental Composition (at. %)
Bulk	DSC	276	269	277	286	8	5.9	21.3		$\text{Ni}_{55.7\pm0.2}\text{Fe}_{11.9\pm0.2}\text{Mn}_{7.2\pm0.1}\text{Ga}_{25.2\pm0.4}$
	$M(T)^{5mT}$	276	273	280	282	6	-	-	301	
SPS	DSC	278	262	279	290	11	5.5	19.1		$\text{Ni}_{55.8\pm0.1}\text{Fe}_{12.0\pm0.1}\text{Mn}_{6.9\pm0.1}\text{Ga}_{25.3\pm0.1}$
	$M(T)^{5mT}$	276	272	280	286	10	-	-	295	

The XRD patterns shown in Figure 2b,c were obtained with an X-ray beam impinging on a plane perpendicular to the thermal gradient during the solidification for the bulk arc-melted sample and on the surface of the sintered disc for the SPS sample, respectively. The present phases in the samples were initially identified based on the ICDD PDF cards 04-014-5690 and 01-077-7907 as a cubic austenite with an L2$_1$ crystal structure and lattice parameter a = 5.774(7) Å (in good agreement with the cooling DSC curves), and a cubic gamma phase with a = 3.614(7) Å) [11], respectively. Vertical dashed lines were drawn in Figure 2b,c across both XRD graphs to underline that for both samples, the respective Bragg reflections of austenite appear in the same 2θ position, whereas the additional weak reflections observed in the patterns correspond to the γ phase. To roughly estimate the amount of gamma phase, a piece of the bulk arc-melted sample was finely pulverized to collect its diffraction pattern. In Figure S1 in the Supplementary Material, the Rietveld refinement of this pattern considering the coexistence of these two phases is shown; the refinement was performed using the Powder Cell program, version 2.4. The amount of γ phase was quantified as ~6% vol. Finally, it must be noted that the peak intensities in the XRD pattern for the SPS sample correspond to an isotropic polycrystal, whereas for the bulk arc-melted sample, austenite grains preferentially grow in the (100) direction oriented along the thermal gradient during solidification. In order to obtain further information about the partial preferential and non-preferential grain orientation in the studied samples, the samples were prepared for SEM observations as follows. The bulk and SPS samples were cut parallel to the thermal gradient during solidification and to the pressing direction during the SPS process, respectively, and carefully polished. Their typical microstructures are shown in Figures 3a and 3b, respectively. The former shows coarse column-shaped grains (note the 2 mm magnification scale) whose major growth direction is parallel to the thermal gradient during solidification (schematically represented in the drawing inserted into Figure 3a). By contrast, the microstructure of the SPS sample is formed by smaller polyhedral grains (<50 μm) not showing a preferential orientation relative to the pressing direction (i.e., the shape of the manually crushed powder particles is replicated in the sintered sample). Hence, the respective XRD patterns confirm the preferential grain orientation of the arc-melted sample as well as the isotropic nature of the SPS sample, while the microstructural analysis verifies the characteristic grain morphology of each synthesis method.

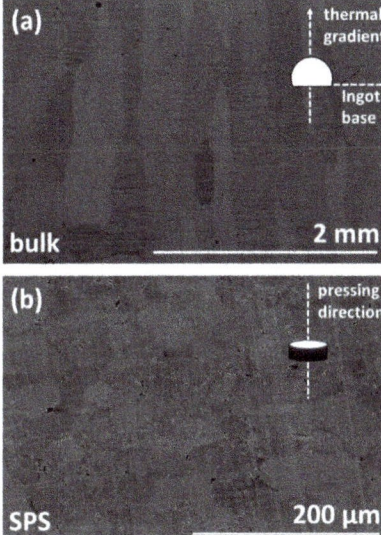

Figure 3. SEM micrographs of the typical microstructure observed for the bulk arc-melted alloy in the plane parallel to thermal gradient (**a**) and for the SPS sample in the plane parallel to pressing direction during the SPS process (**b**). The white solid semi-circle and the vertical dashed arrow in (**a**) schematically represent the cross-sectional view of the arc-melted ingot with its nearly flat bottom surface and the direction of thermal gradient during solidification, respectively. The white vertical dashed line and elliptical area in (**b**) represent the pressing direction during SPS process and the analyzed plane, correspondingly.

$M(T)$ curves under 5 mT and 5 T following zero-field-cooling (ZFC) and field-cooling (FC) regimens in the structural phase transition region are depicted in Figure 4a,b. From the curves measured at 5 mT, the characteristic structural transition temperatures were also determined by simple extrapolation, whereas the temperature of austenite T_C^A was obtained from the minimum of the $dM/dT(T)$ curves; the results are listed in Table 1. Notably, there is good agreement between the values determined for the structural transition temperatures from the $M(T)^{5mT}$ curves and the DSC scans; the observable slight 6 K difference in T_C^A was attributed to the different last thermal treatment that they received. Furthermore, a positive shift in the transformation temperatures ($dT_M/d\mu_0 H$) of 0.4 KT^{-1} was estimated in both the bulk and SPS samples.

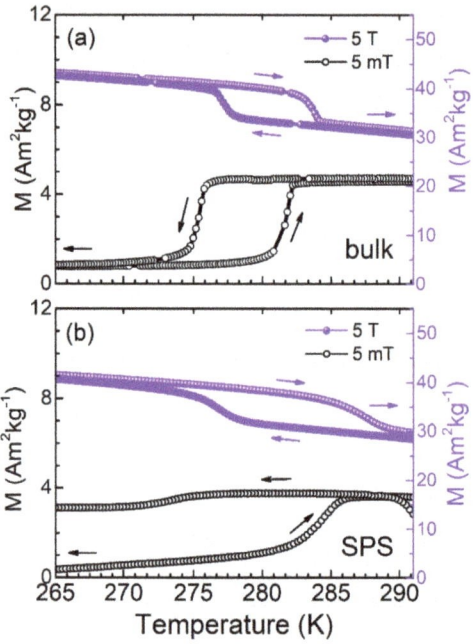

Figure 4. ZFC and FC $M(T)$ curves under 5 mT and 5 T in the temperature range where the structural phase transition occurs for bulk (**a**) and SPS (**b**) samples. The horizontal arrows indicate the scale that correspond to the $M(T)$ values measured at 5 mT (**left**) and 5 T (**right**), respectively.

The magnetic entropy change as a function of the temperature $\Delta S_M(T)$ curves through direct and inverse MT were estimated from the numerical integration of the Maxwell relation (i.e., $\Delta S_M(T, \mu_0 \Delta H) = \mu_0 \int_0^{H_{max}} \left[\frac{\partial M(T,H')}{\partial T}\right]_{H'} dH'$). The sets of isofield $M(T)$ curves measured for this purpose are depicted in Figure 5a,b,c,d,e; they were measured by always crossing the corresponding phase transition in the same direction (up to a maximum applied magnetic field $\mu_0 H_{max}$ of 5 T). The positive shift in the transition underlines the occurrence of a conventional magnetocaloric effect. It must be noted that the magnetization change across the magnetostructural transition is similar in both samples (~6.5 Am^{2}kg^{-1}), but the transition is sharp in the arc-melted sample, suggesting a higher $|\Delta S_M|^{max}$ (due to its direct proportionality to $|dM/dT|$). This also explains the different full width at half-maximum temperature interval δT_{FWHM} of the respective $\Delta S_M(T)$ curves (as shown in Figure 5c,f). Bulk sample exhibits a $|\Delta S_M|^{max}$ of 18.6 and 14.7 Jkg^{-1}K^{-1} for direct and inverse MT, respectively, under $\mu_0 \Delta H_{max}$ = 5 T (δT_{FWHM} = 2.0 and 2.2 K). The difference observed in $|\Delta S_M|^{max}$ between the cooling and heating paths is attributed to the fact that: (i) the formation of austenite and martesite follow different mechanisms and (ii) during cooling, the thermal range in which the phase induction takes place is smaller. By contrast, the SPS sample displays $|\Delta S_M|^{max}$ values of 8.5 in both directions (δT_{FWHM} = 4.0 and 4.4 K). In this case, the similar $|\Delta S_M|^{max}$ value obtained for heating and cooling transitions could be related to the smaller average grain size of the sample, which limits the effect of the phase induction mechanisms. However, what seems to be more decisive for this sample is that the thermal range of the transition for both cooling and heating processes is similar. Despite the $|\Delta S_M|^{max}$ difference, the relative cooling power, RCP, which measures the amount of energy that it is possible to exchange per cooling cycle under ideal conditions, results in 28 Jkg^{-1} at 5 T for both samples. The RCP was obtained by calculating the area below the $|\Delta S_M(T)|$ curves in the temperature interval δT_{FWHM}. It is worth highlighting that despite the moderate values of $|\Delta S_M|^{max}$ displayed, they exceed the largest values previously reported for the studied alloy system ($|\Delta S_M|^{max}$ = 7.5–12.0 Jkg^{-1}K^{-1}) [15]. The magnetocaloric parameters for both the direct and reverse MT for both samples are summarized in Table 2.

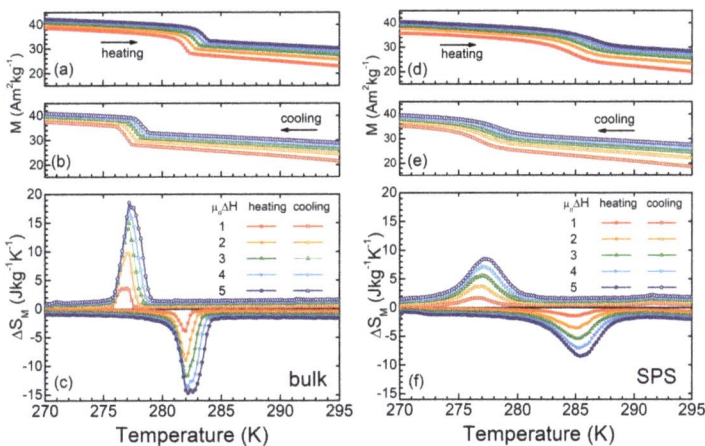

Figure 5. Left and right columns present results for bulk and SPS samples, respectively. The sections (a,b) and (d,e) show the isofield $M(T)$ curves under fields from 1 to 5 T measured across the reverse and direct MT, respectively. The sections (c,f) show the $\Delta S_M(T)$ curves through the magnetostructural transition in both directions.

Table 2. Magnetocaloric properties for a magnetic field change $\mu_0 \Delta H_{max}$ = 5 T for bulk arc-melted and SPS samples through direct and reverse martensitic transformation.

Sample	Structural Transition	$\|\Delta S_M\|^{max}$ (Jkg^{-1}K^{-1})	δT_{FWHM} (K)	RCP (Jkg^{-1})
Bulk	Direct MT	18.6	2.0	28
	Inverse MT	14.7	2.2	
SPS	Direct MT	8.5	4.0	28
	Inverse MT	8.5	4.4	

To characterize the eC effect, direct measurements of ΔT_{ad}^{me} were performed through successive loading and unloading cycles of uniaxial compressive stress σ. Figure 6a shows the time dependence of ΔT_{ad}^{me} at 297 K with a loading and unloading stress of 120 MPa. Figure 6b displays the adiabatic temperature change dependence on temperature, i.e., the $\Delta T_{ad}^{me}(T)$ curves determined for both samples loading/unloading an uniaxial strain of 120 MPa. The bulk sample exhibits a $|\Delta T_{ad}^{me}|^{max}$ of 2.4 K after stress removal, whereas this magnitude significantly decreases to 0.8 K in the SPS sample. This result agrees with the near-to-one-third reduction of the eC effect reported between the directions (100) and

(110) for two MSMAs, such as a single crystal of $Ni_{50}Fe_{19}Ga_{27}Co_4$ [32] and a partially textured polycrystalline arc-melted $Ni_{50}Mn_{32}In_{16}Cr_2$ alloy [33]. A remarkable texture-induced enhancement of the elastocaloric effect was also recently reported for the Mn-rich polycrystalline $Ni_{44}Mn_{46}Sn_{10}$ alloy obtained by directional solidification [34]. Hence, for the magnetocaloric effect, the above-mentioned partial crystallographic texture explains the superior eC effect that the bulk arc-melted sample shows. The observed result also agrees with the fact that in these alloys, the Schmid factor to produce shear is larger in the (100) direction than in the (110) direction. Since the martensite phase is easier to induce in the (100) direction, the $\Delta T_{ad}^{me}(T)$ curves for the bulk arc-melted sample are broader than those of the SPS sample (35 K versus 5 K). The relationship between the $|\Delta T_{ad}^{me}|^{max}$ and the magnitude of the applied stress $\Delta\sigma$, i.e., $|\Delta T_{ad}^{me}|^{max}/\Delta\sigma$ ratio obtained for the bulk and SPS samples are 21 kGPa^{-1} and 7 kGPa^{-1}, respectively. For the bulk sample, it is comparable with the value reported in other polycrystalline alloys [14]. The $\Delta T_{ad}^{me}/\Delta\sigma$ ratio in the SPS sample is also compatible with that measured in a Ni-Fe-Ga-Co (011)-oriented single crystal [35].

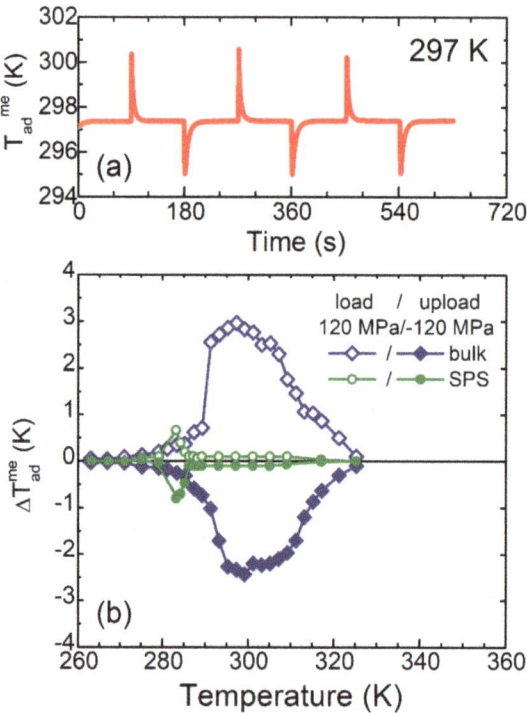

Figure 6. ΔT_{ad}^{me} as a function of time at 297 K for the bulk sample (**a**) and $\Delta T_{ad}^{me}(T)$ for both samples under loaded and unload applied stress of 120 MPa (**b**).

4. Conclusions

In summary, from the study carried out on the structural, microstructural, and phase transition characteristics, as well as the magnetocaloric and elastocaloric effects linked to the martensitic-like transformation in bulk samples of the magnetic-shaped memory alloy $Ni_{55}Fe_{11}Mn_7Ga_{27}$ produced by arc melting and spark plasma sintering, the following was demonstrated: (a) the effectiveness of the spark plasma sintering technique at consolidating highly dense polycrystalline samples of (Ni-Fe-Ga)-based alloys without showing significant changes in phase transformation temperature or in crystal structure and displaying similar MC properties compared with those obtained from the bulk sample. As far as we know, this is the first report about the consolidation of these alloys by means of this processing technique; (b) the significant role of crystallographic texture in the enhancement of elastocaloric responses linked to the first-order structural transformation in Mn-doped (Ni-Fe-Ga)-based alloys.

Supplementary Materials: The following supporting information can be downloaded at: https://www.mdpi.com/article/10.3390/met12020273/s1, Figure S1: Rietveld refinement of the X-ray powder diffraction pattern of a finely pulverized sample piece taken from the bulk arc melted sample. The black open circles and red solid line represent the experimental points and Rietveld refined data, respectively, whereas the difference between experimental and simulated patterns is given by the blue line depicted at the bottom of the figure. The R-factor values obtained for the refinement were: $R_p = 11.25\%$, $R_{wp} = 28.12\%$ and $R_{exp} = 2.38\%$.

Author Contributions: Conceptualization, H.F.-Z. and J.P.C.-G.; software, J.D.N.-G. and J.P.C.-G.; validation, F.A.-H., H.F.-Z. and J.L.S.L.; formal analysis, F.A.-H.; investigation, J.D.N.-G., J.P.C.-G. and J.L.S.L; writing—original draft preparation, J.D.N.-G. and J.P.C.-G.; writing—review and editing, J.L.S.L.; supervision, J.L.S.L., H.F.-Z. and J.P.C.-G. All authors have read and agreed to the published version of the manuscript.

Funding: This research was funded by SEP-CONACyT, grant number A1-S-37066. J.D. Navarro-García and J.P. Camarillo-Garcia are grateful to CONACyT-Mexico for supporting their Ph.D. studies at IPICyT and postdoctoral position at UAZ, respectively.

Data Availability Statement: The data that supports the findings of this study are available within the article.

Acknowledgments: This work was supported by SEP-CONACyT, Mexico (under the research project A1-S-37066) and Laboratorio Nacional de Nanociencias y Nanotecnología (LINAN, IPICyT), where most of the experimental work was carried out. The authors are grateful to B. A. Rivera Escoto and A.I. Peña Maldonado for their technical support. J.D. Navarro-García and J.P. Camarillo-Garcia are grateful to CONACyT-Mexico for supporting their Ph.D. studies at IPICyT and postdoctoral position at UAZ, respectively.

Conflicts of Interest: The authors declare that they have no known competing financial interests or personal relationships that could have appeared to influence the work reported in this paper.

References

1. Mañosa, L.; Planes, A. Materials with giant mechanocaloric effects: Cooling by strength. *Adv. Mater.* **2017**, *29*, 1603607. [CrossRef] [PubMed]
2. Castillo-Villa, P.O.; Soto-Parra, D.E.; Matutes-Aquino, J.A.; Ochoa-Gamboa, R.A.; Planes, A.; Mañosa, L.; González-Alonso, D.; Stipcich, M.; Romero, R.; Ríos-Jara, D.; et al. Caloric effects induced by magnetic and mechanical fields in a $Ni_{50}Mn_{25-x}Ga_{25}Co_x$ magnetic shape memory alloy. *Phys. Rev. B* **2011**, *83*, 174109. [CrossRef]
3. Li, Y.; Zhao, D.; Liu, J. Giant and reversible room-temperature elastocaloric effect in a single-crystalline Ni-Fe-Ga magnetic shape memory alloy. *Sci. Rep.* **2016**, *6*, 25500. [CrossRef]
4. Bonnot, E.; Romero, R.; Mañosa, L.; Vives, E.; Planes, A. Elastocaloric effect associated with the martensitic transition in shape-memory alloys. *Phys. Rev. Lett.* **2008**, *100*, 125901. [CrossRef] [PubMed]
5. Yan, H.-L.; Wang, L.-D.; Liu, H.X.; Huang, X.M.; Jia, N.; Li, Z.B.; Yang, B.; Zhang, Y.D.; Esling, C.; Zhao, X.; et al. Giant elastocaloric effect and exceptional mechanical properties in an all-d-metal Ni–Mn–Ti alloy: Experimental and ab-initio studies. *Mater. Des.* **2019**, *184*, 108180. [CrossRef]
6. Chauhan, A.; Patel, S.; Vaish, R.; Bowen, C.R. A review and analysis of the elasto-caloric effect for solid-state refrigeration devices: Challenges and opportunities. *MRS Energy Sustain.* **2015**, *2*, E16. [CrossRef]
7. Franco, V.; Blázquez, J.S.; Ipus, J.J.; Law, J.Y.; Moreno-Ramírez, L.M.; Conde, A. Magnetocaloric effect: From materials research to refrigeration devices. *Prog. Mater. Sci.* **2018**, *93*, 112–232. [CrossRef]
8. Castillo-Villa, P.O.; Mañosa, L.; Planes, A.; Soto-Parra, D.E.; Sánchez-Llamazares, J.L.; Flores-Zúñiga, H.; Frontera, C. Elastocaloric and magnetocaloric effects in Ni-Mn-Sn(Cu) shape-memory alloy. *J. Appl. Phys.* **2013**, *113*, 053506. [CrossRef]
9. Mañosa, L.; Planes, A. Mechanocaloric effects in shape memory alloys. *Philos. Trans. R. Soc. A* **2016**, *374*, 20150310. [CrossRef] [PubMed]
10. Oikawa, K.; Ōmori, T.; Sutou, Y.; Morito, H.; Kainuma, R.; Ishida, K. Phase equilibria and phase transition of the Ni–Fe–Ga ferromagnetic shape memory alloy System. *Metall. Mater. Trans. A* **2007**, *38*, 767–776. [CrossRef]

11. Omori, T.; Kamiya, N.; Sutou, Y.; Oikawa, K.; Kainuma, R.; Ishida, K. Phase transformations in Ni-Ga-Fe ferromagnetic shape memory alloys. *Mater. Sci. Eng. A.* **2004**, *378*, 403–408. [CrossRef]
12. Li, J.Q.; Liu, Z.H.; Yu, H.C.; Zhang, M.; Zhou, Y.Q.; Wu, G.H. Martensitic transition and structural modulations in the Heusler alloy Ni_2FeGa. *Solid State Commun.* **2003**, *126*, 323–327. [CrossRef]
13. Jiang, C.; Muhammad, Y.; Deng, L.; Wu, W.; Xu, H. Composition dependence on the martensitic structures of the Mn-rich NiMnGa alloys. *Acta Mater.* **2004**, *52*, 2779–2785. [CrossRef]
14. Xu, Y.; Lu, B.; Sun, W.; Yan, A.; Liu, J. Large and reversible elastocaloric effect in dual-phase $Ni54Fe19Ga27$ superelastic alloys. *Appl. Phys. Lett.* **2015**, *106*, 201903. [CrossRef]
15. Sarkar, S.K.; Biswas, A.; Babu, P.D.; Kaushik, S.D.; Srivastava, A.; Siruguri, V.; Krishnan, M. Effect of partial substitution of Fe by Mn in $Ni_{55}Fe_{19}Ga_{26}$ on its microstructure and magnetic properties. *J. Alloys Compd.* **2014**, *586*, 515–523. [CrossRef]
16. Yu, H.J.; Zu, X.T.; Fu, H.; Zhang, X.Y.; Wang, Z.G. Effect of annealing and heating/cooling rate on the transformation temperatures of NiFeGa alloy. *J. Alloys Compd.* **2009**, *470*, 237–240. [CrossRef]
17. Tian, X.H.; Sui, J.H.; Zhang, X.; Zheng, X.H.; Cai, W. Grain size effect on martensitic transformation. mechanical and magnetic properties of Ni-Mn-Ga alloy fabricated by spark plasma sintering. *J. Alloys Compd.* **2012**, *514*, 210–213. [CrossRef]
18. Bai, J.; Liu, D.; Gu, J.; Jiang, X.; Liang, X.; Zhang, Y.; Zhao, X. Excellent mechanical properties and large magnetocaloric effect of spark plasma sintered Ni-Mn-In-Co alloy. *J. Mater. Sci. Technol.* **2021**, *74*, 46–51. [CrossRef]
19. Imam, H.; Zhang, H.G.; Chen, J.; Yue, M.; Lu, Q.M.; Zhang, D.T.; Liu, W.Q. Powdering and SPS sintering effect on the magnetocaloric properties of MnNiSi-based compounds. *AIP Adv.* **2019**, *9*, 035205. [CrossRef]
20. Kuang, Y.; Ai, Z.; Yang, B.; Hao, X.; Li, Z.; Yan, H.; Zhang, Y.; Esling, C.; Zhao, X.; Zuo, L. Simultaneously achieved good mechanical properties and large magnetocaloric effect in spark plasma sintered Ni-Mn-In alloys. *Intermetallics* **2020**, *124*, 106868. [CrossRef]
21. Ito, K.; Ito, W.; Umetsu, R.Y.; Tajima, S.; Kawaura, H.; Kainuma, R.; Ishida, K. Metamagnetic shape memory effect in polycrystalline NiCoMnSn alloy fabricated by spark plasma sintering. *Scr. Mater.* **2009**, *61*, 504–507. [CrossRef]
22. Li, Y.; Jiang, C.; Liang, T.; Ma, Y.; Xu, H. Martensitic transformation and magnetization of Ni-Fe-Ga ferromagnetic shape memory alloys. *Scr. Mater.* **2003**, *48*, 1255–1258. [CrossRef]
23. Barandiarán, J.M.; Gutiérrez, J.; Lázpita, P.; Chernenko, V.A.; Seguí, C.; Pons, J.; Cesari, E.; Oikawa, K.; Kanomata, T. Martensitic transformation in Ni-Fe-Ga alloys. *Mater. Sci. Eng. A* **2008**, *478*, 125–129. [CrossRef]
24. Kustov, S.; Corró, M.; Cesari, E. Stress-induced magnetization in polycrystalline Ni-Fe-Ga ferromagnetic shape memory alloy. *Appl. Phys. Lett.* **2007**, *91*, 141907. [CrossRef]
25. Pérez-Landazábal, J.I.; Recarte, V.; Gómez-Polo, C.; Seguí, C.; Cesari, E.; Dutkiewicz, J. Magnetic behavior in Ni-Fe-Ga martensitic phase. *Mater. Sci. Eng. A* **2008**, *481–482*, 318–321. [CrossRef]
26. Liu, Z.H.; Liu, H.; Zhang, X.X.; Zhang, M.; Dai, X.F.; Hu, H.N.; Chen, J.L.; Wu, G.H. Martensitic transformation and magnetic properties of Heusler alloy Ni-Fe-Ga ribbon. *Phys. Lett. A* **2004**, *329*, 214–220. [CrossRef]
27. Recarte, V.; Pérez-Landazábal, J.I.; Gómez-Polo, C.; Cesari, E.; Dutkiewicz, J. Magnetocaloric effect in Ni-Fe-Ga shape memory alloys. *Appl. Phys. Lett.* **2006**, *88*, 132503. [CrossRef]
28. Bruno, N.M.; Karaman, I.; Chumlyakov, Y.I. Orientation dependence of the elastocaloric effect in $Ni_{54}Fe_{19}Ga_{27}$ ferromagnetic shape memory alloy. *Phys. Status Solidi B* **2018**, *255*, 1700437. [CrossRef]
29. Guillon, O.; Gonzalez-Julian, J.; Dargatz, B.; Kessel, T.; Schierning, G.; Räthel, J.; Herrmann, M. Field-assisted sintering technology/spark plasma sintering: Mechanisms. materials, and technology developments. *Adv. Eng. Mater.* **2014**, *16*, 830–849. [CrossRef]
30. Bruno, N.M. The Magnetocaloric and Elastocaloric Effects in Magnetic Shape Memory Alloys. Ph.D. Thesis, Texas A&M University, College Station, TX, USA, 2015.
31. Marcos, J.; Planes, A.; Mañosa, L. Magnetic field induced entropy change and magnetoelasticity in Ni-Mn-Ga alloys. *Phys. Rev. B* **2002**, *66*, 224413. [CrossRef]
32. Xiao, F.; Jin, M.; Liu, J.; Jin, X. Elastocaloric effect in $Ni_{50}Fe_{19}Ga_{27}Co_4$ single crystals. *Acta Mater.* **2015**, *96*, 292–300. [CrossRef]
33. Henández-Navarro, F.; Camarillo-Garcia, J.P.; Aguilar-Ortiz, C.O.; Flores-Zúñiga, H.; Ríos, D.; González, J.G.; Álvarez-Alonso, P. The influence of texture on the reversible elastocaloric effect of a polycrystalline Ni50Mn32In16Cr2 alloy. *Appl. Phys. Lett.* **2018**, *112*, 164101. [CrossRef]
34. Zhang, G.; Li, Z.; Yang, J.; Yang, B.; Wang, D.; Zhang, Y.; Esling, C.; Hou, L.; Li, X.; Zhao, X.; et al. Giant elastocaloric effect in a Mn-rich $Ni_{44}Mn_{46}Sn10$ directionally solidified alloy. *Appl. Phys. Lett.* **2020**, *116*, 023902. [CrossRef]
35. Zhao, D.; Xiao, F.; Nie, Z.; Cong, D.; Sun, W.; Liu, J. Burst-like superelasticity and elastocaloric effect in [011] oriented $Ni_{50}Fe_{19}Ga_{27}Co_4$ single crystals. *Scr. Mater.* **2018**, *149*, 6–10. [CrossRef]

Article

Characterizing Changes in Grain Growth, Mechanical Properties, and Transformation Properties in Differently Sintered and Annealed Binder-Jet 3D Printed 14M Ni–Mn–Ga Magnetic Shape Memory Alloys

Aaron Acierno [1], Amir Mostafaei [1,†], Jakub Toman [1], Katerina Kimes [1], Mirko Boin [2], Robert C. Wimpory [2], Ville Laitinen [3], Andrey Saren [3], Kari Ullakko [3] and Markus Chmielus [1,*]

[1] Department of Mechanical Engineering and Materials Science and Engineering, University of Pittsburgh, Pittsburgh, PA 15261, USA; ama195@pitt.edu (A.A.); mostafaei@iit.edu (A.M.); jakub.toman@pitt.edu (J.T.); kak272@pitt.edu (K.K.)

[2] Helmholtz-Zentrum Berlin für Materialien und Energie GmbH, Hahn-Meitner-Platz 1, 14109 Berlin, Germany; boin@helmholtz-berlin.de (M.B.); robert.wimpory@helmholtz-berlin.de (R.C.W.)

[3] School of Engineering Science, Lappeenranta-Lahti University of Technology LUT, 53850 Lappeenranta, Finland; ville.laitinen@lut.fi (V.L.); andrey.saren@lut.fi (A.S.); kari.ullakko@lut.fi (K.U.)

* Correspondence: chmielus@pitt.edu

† Current address: Department of Mechanical, Materials and Aerospace Engineering, Illinois Institute of Technology, Chicago, IL 60616, USA.

Citation: Acierno, A.; Mostafaei, A.; Toman, J.; Kimes, K.; Boin, M.; Wimpory, R.C.; Laitinen, V.; Saren, A.; Ullakko, K.; Chmielus, M. Characterizing Changes in Grain Growth, Mechanical Properties, and Transformation Properties in Differently Sintered and Annealed Binder-Jet 3D Printed 14M Ni–Mn–Ga Magnetic Shape Memory Alloys. *Metals* **2022**, *12*, 724. https://doi.org/10.3390/met12050724

Academic Editor: João Pedro Oliveira

Received: 20 March 2022
Accepted: 20 April 2022
Published: 24 April 2022

Publisher's Note: MDPI stays neutral with regard to jurisdictional claims in published maps and institutional affiliations.

Copyright: © 2022 by the authors. Licensee MDPI, Basel, Switzerland. This article is an open access article distributed under the terms and conditions of the Creative Commons Attribution (CC BY) license (https://creativecommons.org/licenses/by/4.0/).

Abstract: Ni–Mn–Ga Heusler alloys are multifunctional materials that demonstrate macroscopic strain under an externally applied magnetic field through the motion of martensite twin boundaries within the microstructure. This study sought to comprehensively characterize the microstructural, mechanical, thermal, and magnetic properties near the solidus in binder-jet 3D printed 14M $Ni_{50}Mn_{30}Ga_{20}$. Neutron diffraction data were analyzed to identify the martensite modulation and observe the grain size evolution in samples sintered at temperatures of 1080 °C and 1090 °C. Large clusters of high neutron-count pixels in samples sintered at 1090 °C were identified, suggesting Bragg diffraction of large grains (near doubling in size) compared to 1080 °C sintered samples. The grain size was confirmed through quantitative stereology of polished surfaces for differently sintered and heat-treated samples. Nanoindentation testing revealed a greater resistance to plasticity and a larger elastic modulus in 1090 °C sintered samples (relative density ~95%) compared to the samples sintered at 1080 °C (relative density ~80%). Martensitic transformation temperatures were lower for samples sintered at 1090 °C than 1080 °C, though a further heat treatment step could be added to tailor the transformation temperature. Microstructurally, twin variants ≤10 µm in width were observed and the presence of magnetic anisotropy was confirmed through magnetic force microscopy. This study indicates that a 10 °C sintering temperature difference can largely affect the microstructure and mechanical properties (including elastic modulus and hardness) while still allowing for the presence of magnetic twin variants in the resulting modulated martensite.

Keywords: additive manufacturing; ferromagnetic; neutron diffraction; microstructure; nanoindentation; sintering

1. Introduction

Ni_2–Mn–Ga-derivative Heusler materials are considered magnetic shape memory alloys (MSMAs) and demonstrate a large mechanical strain under an externally applied magnetic field [1,2]. This magnetic-field induced strain (MFIS) occurs through twin variant reorientation in the low-temperature ferromagnetic martensite phase, due in part to its large magnetocrystalline anisotropy [2,3]. Ni–Mn–Ga MSMAs were shown to deform by up to 12% with a fast response time in the order of a few microseconds, high working frequencies up to 100 kHz, and long lifetimes exceeding 10^7 cycles [4–9]. These properties have generated interest in alternative materials for use as actuators and sensors from the nano- to macro-scale [10–12]. Additionally, when the Curie and phase transformation temperatures overlap, Ni–Mn–Ga(–X) alloys were shown to exhibit a magnetocaloric effect useful in refrigeration, sensors, and energy harvesters [10,12,13].

Significant MFIS requires a compliant microstructure for the motion of martensite twin boundaries. In particular, the grain size, present phases, compositional additions, and processing and fabrication conditions are engineering choices that must be considered when designing MSMAs exhibiting MFIS. For example, Lázpita et al. [14] showed that fine-grained and randomly textured Ni–Mn–Ga samples exhibit hardly any MFIS under an external magnetic field due to internal microstructural constraints on the twin-boundary motion. Furthermore, surface defects strongly influence twin-boundary stresses and can stabilize fine twin-boundary structures [15–19]. This places a particular emphasis on the processing and fabrication of Ni–Mn–Ga alloys aiming to maximize the MFIS by producing a microstructure with fewer defects, grain boundaries, and interstitials.

To promote MFIS in polycrystalline MSMAs, there has been increasing interest in manufacturing porous Ni–Mn–Ga polycrystalline samples [20]. Unlike fully-densified polycrystalline samples containing a large surface area of grain boundaries, porous Ni–Mn–Ga structures reduce constraints and allow the martensitic twins to move more freely, decreasing the activation stress required to achieve motion [5,9,14,20,21]. A proposed method for manufacturing samples with intentional porosity is through additive manufacturing, where single layers of material are deposited and bound one layer at a time [22]. So far, Taylor et al. [23] have performed work on 3D printed inks using elemental powders, and Mostafaei et al. [24,25] and Caputo et al. [26] have discussed the use of binder-jet 3D printing (or binder jetting) as a fast and cost-effective additive manufacturing technique with the capability to produce complex shapes with tunable porosity. Post-processing sintering and heat-treatment allow user-controlled porosity levels to be achieved following printing through various parameters including sintering temperature, time, and environment [27].

In this study, binder-jet 3D printed $Ni_{50}Mn_{30}Ga_{20}$ (at.%) samples were sintered at two different temperatures and additionally annealed (four samples in total). Microstructural characteristics including lattice parameters, martensite modulation, and unit cell size were correlated to grain size, porosity, and hardness measurements. Properties including martensitic transformation temperature, surface magnetic structure, and magnetization behavior were correlated to post-processing parameters.

2. Materials and Methods

Sample preparation procedures for materials used in this experiment are provided in detail in [24] with a brief summary given here. Polycrystalline ingots were fabricated using induction melting on high-purity elemental Ni, Mn, and Ga powder. Using energy-dispersive X-ray spectroscopy (EDS), the composition of the ingot was determined to be $Ni_{49.7\pm0.5}Mn_{30.0\pm1.0}Ga_{20.3\pm0.6}$ (at.%). The ingots were broken up, ball-milled using a planetary mill (Retsch Inc., Haan, Germany), and sieved using a US 230 ASTM E-11 standard mesh sieve (Fisherbrand, Pittsburgh, PA, USA). The powder was then binder-jet 3D printed using an ExOne X1-Lab printer (The ExOne Company, Irwin, PA, USA). Printing parameters for this fabrication step are as follows: layer thickness of 100 μm, spread speed of 20 mm/s, feed:build powder ratio of 2, drying time of 40 s, and binder saturation of 80%. Additionally, powder layers were bound with an ethylene glycol monomethyl ether and diethylene glycol solvent binder. As-printed green samples were cured at 200 °C for 8 h in air then encapsulated in a glass tube with high purity titanium sponge (as a sacrificial oxidizer) under an argon atmosphere of approximately 40 kPa at room temperature. Encapsulated samples were heated at 4 °C/min to the sintering temperature of 1080 °C or 1090 °C using a Lindberg tube furnace (Lindberg Blue, Asheville, NC, USA), held at this temperature for 2 h, then air-cooled. Additionally, a heat-treatment procedure was performed on one sample of each sintering temperature, which comprised of an annealing step at 1000 °C for 10 h and an ordering step at 700 °C for 12 h, for a total of four samples under investigation.

Neutron diffraction data were obtained for Ni–Mn–Ga samples from the E3 beamline (wavelength λ = 1.47318 ± 0.00056 Å) at the Helmholtz-Zentrum Berlin für Materialien und Energie (HZB) [28]. The beamline utilized a bent Si (400) focusing monochromator and a 30 cm × 30 cm position-sensitive detector (PSD). The detector center was positioned at 2θ values of 42°, 53°, 64°, and 75° while Ω (sample stage, see Figure 1) was rotated in increments of 2° from −90° to 0°, and the intensity on the detector was recorded. A schematic of the goniometer stationed at beamline E3 is shown in Figure 1 with relevant angles illustrated [29]. Area detector raw were was extracted utilizing in-house plotting software, and Gaussian peaks were fit to the diffraction curves using OriginPro software (2018b, OriginLab, Northampton, MA, USA).

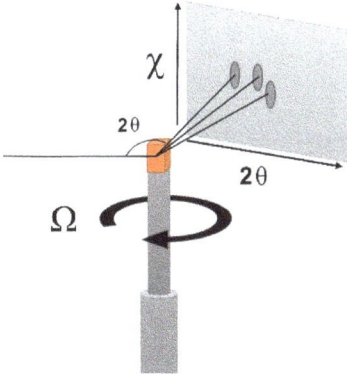

Figure 1. Schematic of the goniometer setup on Beamline E3 at HZB.

Raw neutron diffraction data were extracted image-by-image within the 2θ = 42° regime using ImageJ image analysis software (NIH, Bethesda, MD, USA) for conversion [30,31] and an in-house MATLAB program (MathWorks Inc., Natick, MA, USA) for processing. For the 256 × 256 pixel array, the program isolated, enhanced, and clarified pixel clusters likely corresponding to individual grains satisfying Bragg diffraction criteria. An exemplary frame highlighting grain clustering of the 1090 °C sintered-only sample in the region 2θ = 75° at Ω = −90° is shown in Figure 2a. For all raw images (i.e., all Ω steps) in the 2θ = 42° regime, pixel clusters were counted following image processing for all four samples to provide qualitative interpretation of the grain size.

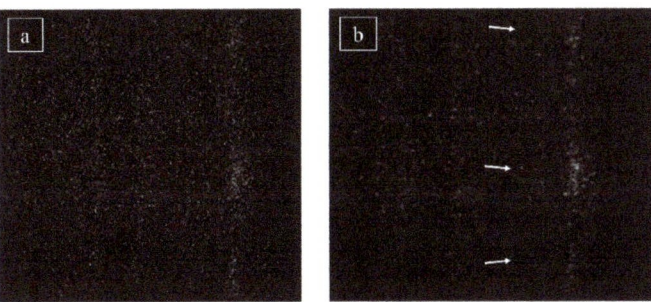

Figure 2. Neutron diffraction area detector image received from HZB before (**a**) and after (**b**) image processing. Exemplary clusters are marked using white arrows.

Samples were mounted, polished, and etched using a Tegramin-25 automatic polisher (Struers, Ballerup, Denmark) and imaged for grain size using a Keyence VHX-600 digital optical microscope (Keyence, Osaka, Japan) with a VH-Z100 lens. Grain sizes from several micrographs were measured and compared using ImageJ image analysis software through quantitative stereology of the etched surfaces utilizing a method previously outlined in [32]. Differential interference contrast (DIC) microscopy was performed using a Nikon Optiphot (Nikon, Scotia, NY, USA) to observe twinning on the sample surfaces.

Mechanical behavior of the samples was quantified through nanoindentation on mounted materials using a Hysitron TI 950 Triboindenter (Bruker, Tuscon, AZ, USA) with a Berkovich indentation tip. The following trapezoidal loading function was used for testing: linear loading from 0 to 10 mN at a rate of 2 mN/s, constant applied load of 10 mN for 5 s, and linear unloading from 10 to 0 mN at a rate of 2 mN/s. Tests were performed at various distances from pores on the sample surface, and the reduced Elastic modulus (E_r) and hardness were recorded in GPa. Ten load–displacement curves were collected and analyzed for each of the four samples to isolate and identify homogenous dislocation motion. Mechanical properties identified via nanoindentation were considered in relation to material densification throughout the sintering process. Relative density of the four samples was determined using Archimedes' principle.

In order to identify first- and second-order major transformation temperatures, differential scanning calorimetry (DSC) was performed using a DSC 250 (TA Instruments, New Castle, DE, USA) on the four investigated samples in a temperature window encapsulating the martensitic phase transformation and Curie temperatures. Cut samples were weighed and ramped at 5 °C/min from 40 °C to 120 °C, held at 120 °C for one minute, and then ramped at 5 °C/min from 120 °C to 40 °C. Major peaks and transformation temperatures were identified, noting the expected presence of a significant thermal hysteresis signature in Ni–Mn–Ga Heusler alloys. On heating, the austenite transformation is fully described by the austenite start (A_S), austenite finish (A_F), and austenite peak (A_P) temperatures. Similarly, the martensite transformation is described by martensite start (M_S), martensite finish (M_F), and martensite peak (M_P) temperatures on cooling. The average transformation temperature (T_{trans}) averaging the martensite and austenite transformation peaks is also reported. Finally, the Curie temperature (T_C) was determined for all samples.

Vibrating sample magnetometry (VSM) was performed using a series 7400 model vibrating sample magnetometer (Lakeshore, Westerville, OH, USA) and magnetization curves in an external magnetic field (± 1.5 T) were analyzed to determine saturation magnetization and magnetic coercivity. Additionally, a set of samples sintered at 1080 °C and 1090 °C (non-heat-treated) were electropolished at a constant voltage of 12 V at -20 °C in an electrolyte solution containing a 3:1 volumetric ratio of ethanol to 60% HNO_3. Atomic force microscopy (AFM) and magnetic force microscopy (MFM) was performed using an Park XE7 atomic force microscope (Park Systems, Suwon, Korea) on areas identified (using polarized light microscopy) as regions containing highly dense twin boundaries.

3. Results and Discussion

McIntyre discussed that it is possible to identify single crystalline regions in a material using neutron diffraction area detector raw data [33]. It is expected that large clusters in various regions of the raw data (e.g., in Figure 2) correspond to a distinct single grain or crystal which reflects all permissible (*hkl*) lattice planes according to the Bragg geometry. In other words, by processing individual detector images and observing grain clusters, the relative size of grains can be qualitatively compared. Figure 3 (bottom row) shows examples of processed single Ω detector images of the $2\theta = 42°$ position. It follows that larger grains should be present in a Ω sweep containing large and coherent clusters. It is observed that the 1090 °C sintered samples have more distinguishable diffraction clusters compared to the 1080 °C sintered samples, which show more continuous Bragg diffraction ring sections.

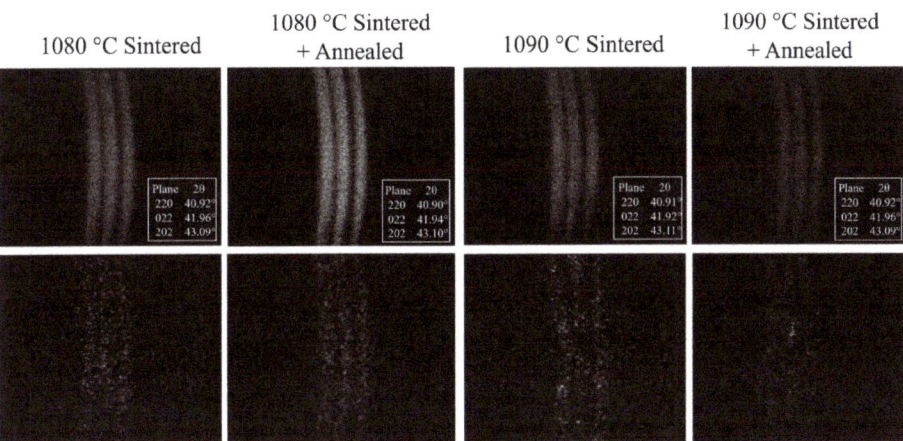

Figure 3. Processed and integrated area detector diffraction images (**top**) and individual example processed diffraction images taken at one Ω step (**bottom**) for each of the four samples. Images are taken of the (220) plane family of 14M Ni–Mn–Ga martensite. Insets of each diffraction image display the location of the integrated diffraction curve for each band. These diffraction angles were used to determine lattice parameters for each sample.

Figure 3 (top row) shows an example of an integrated detector image following the processing described prior. The integrated diffraction images at the $2\theta = 42°$ position were used to identify the martensite phase at room temperature. These three rings correspond to the (220) lattice plane family of 14M martensite for all four samples. The lattice parameters were calculated using the (400) and (220) lattice plane families using a cubic Cartesian coordinate system with an assumed orthorhombic unit cell ($\alpha = \beta = \gamma = 90°$). The c/a ratio is similar for all samples (between 0.8942 and 0.8965) and corresponds to a 14M martensitic structure [24]. A summary of the lattice parameters determined by neutron diffraction, including c/a ratio and unit cell volume, is shown in Table 1.

Table 1. Crystallographic, mechanical, physical, and phase transformation properties of the Ni–Mn–Ga binder jetted samples.

Parameter	1080 °C Sintered	1080 °C Sintered and Annealed	1090 °C Sintered	1090 °C Sintered and Annealed
a (Å)	6.169 ± 0.004	6.169 ± 0.005	6.190 ± 0.005	6.177 ± 0.018
b (Å)	5.822 ± 0.003	5.833 ± 0.004	5.820 ± 0.003	5.823 ± 0.016
c (Å)	5.531 ± >0.001	5.530 ± >0.001	5.526 ± 0.001	5.530 ± >0.001
c/a	0.8965	0.8958	0.8942	0.8958
Unit Cell Volume (Å^3)	198.6 ± 0.2	198.9 ± 0.3	199.0 ± 0.5	198.9 ± 1.1
Hardness (GPa)	3.00 ± 0.24	3.36 ± 0.37	3.47 ± 0.24	3.49 ± 0.22
Reduced Elastic Modulus (GPa)	92.6 ± 7.3	88.3 ± 13.7	104.9 ± 12.2	108.5 ± 7.0
Relative Density (%)	79.4 ± 5.6	80.1 ± 2.0	94.2 ± 2.3	94.6 ± 3.5

Table 1. Cont.

Parameter	1080 °C Sintered	1080 °C Sintered and Annealed	1090 °C Sintered	1090 °C Sintered and Annealed
Foam Elastic Modulus (GPa)	91.7 ± 7.97	87.1 ± 14.7	105 ± 13.8	109 ± 7.90
Projected Bulk Elastic Modulus (GPa)	147 ± 11.7	136 ± 23.0	119 ± 15.6	123 ± 8.98
Grain Diameter (μm)	43.6 ± 1.9	48.9 ± 2.9	82.4 ± 2.6	91.6 ± 2.4
A_S (°C)	84.5	82.4	75.6	80.0
A_F (°C)	91.2	91.3	82.5	89.9
A_P (°C)	89.0	89.2	79.3	86.3
M_S (°C)	86.7	87.2	77.1	82.0
M_F (°C)	78.8	75.7	68.7	70.2
M_P (°C)	83.5	83.2	72.4	77.4
T_{trans} (°C)	86.3	86.2	75.9	81.9
T_C (°C)	88	88	88	87
M_{sat} (Am^2/kg)	52.4	56.4	58.0	57.9

A qualitative grain size comparison by counting diffraction area clusters was conducted for all samples using an entire Ω sweep (46 images at 2° increments) in the 2θ = 42° regime. Each Ω image was processed as described above, and distinct and coherent clusters in the area detector diffraction images through subjective human interpretation. Through this procedure, nearly three times the number of distinct clusters were identified in the 1090 °C sintered samples (nearly 70 counted for each) compared to the 1080 °C sintered samples (nearly 25 counted for each). The large difference in counted clusters suggests that the 10 °C increase in sintering temperature has more than doubled the number of crystallites large enough to be identifiable via neutron diffraction and, thus, has a measurable impact on the final grain size. It was also determined that the optional heat-treatment step used here does not contribute significantly to grain growth. Overholser et al. [34] reported a solidus temperature of 1088 °C for samples of composition $Ni_{50}Mn_{30}Ga_{20}$, suggesting the possibility of at least partial super-solidus liquid phase sintering of the binder-jet 3D printed Ni–Mn–Ga samples sintered at 1090 °C. Mostafaei et al. [24] identified similar sintering behavior by investigating the sintering characteristics of binder-jet 3D printed Ni–Mn–Ga at various temperatures near the solidus. This is observed to coincide with massive densification (~80% to ~95%) of the microstructure through the 10 °C sintering temperature increase, as seen in Table 1. See Supplementary Figure S1 for example SEM images of observable grain boundaries on the material microstructure and Supplementary Figure S2 of a micrograph of an etched sample with marked grain boundaries.

Optical microscopy was applied to visually identify grain boundaries and distinguish individual grains on various cross-sectional areas of magnification identical for all samples (see Table 1 for average equivalent grain diameter). This quantitative approach to grain size identification provides that a sintering temperature increase of 10 °C almost doubles the grain diameter. This confirms the qualitative interpretation that the increase in visible clusters on neutron diffraction images through increased sintering temperature increases grain sizes while the additional heat-treatment shows little effect. Additionally, DIC micrographs for all four samples (presented in Figure 4) show fine twins that are primarily ≤10 μm in width, with some twins showing a width of approximately 1–2 μm. Twin boundaries, which separate the twin variants within crystallite regions, are a requirement for MFIS. Larger grains will contain less grain boundary surface area per volume, reducing twin-boundary motion incompatibilities. Although the literature has discussed the introduction of intentional porosity to reduce the internal constraints on the twin variant reorientation, large densification observed by sintering at 1090 °C largely purges these intentional pores. Chmielus et al. [21] demonstrated large MFIS (up to 8.7%) for polycrystalline foams with pores smaller than the average grain size. Clearly, both grain size and porosity percentage can be adjusted through the introduction of a sintering step, however, the interplay between grain size and pore sizes should be explored as a function of sintering temperature to further optimize MFIS. Porosity characterization could be explored using micro-computed X-ray tomography to identify and analyze pore channels and isolated pore features and dimensions (utilized by our group previously to determine relative density in [24]).

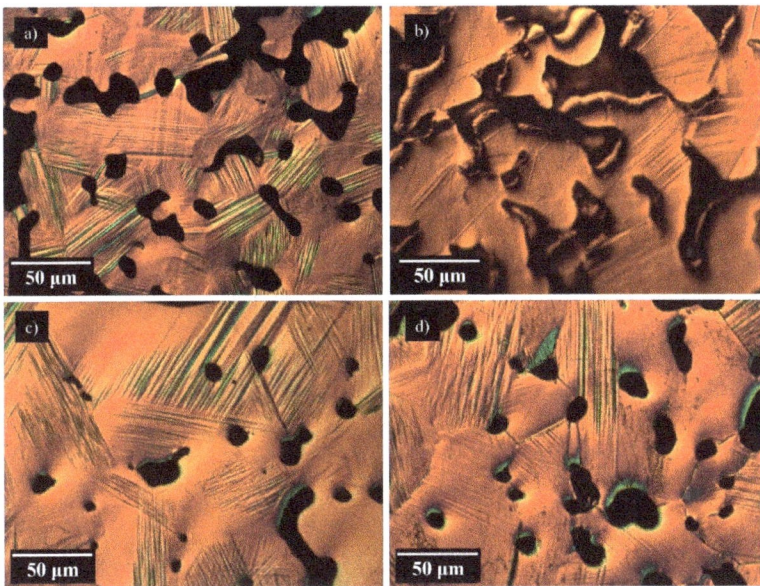

Figure 4. DIC micrographs of polished samples for the investigated materials during this study. The top row presents the 1080 °C sintered sample (**a**) and 1080 °C sintered + heat-treated sample (**b**), respectively, whereas the bottom row presents the 1090 °C sintered sample (**c**) and 1090 °C sintered + heat-treated sample (**d**).

Nanoindentation results, summarized in Table 1, show that among the four samples, the 1080 °C non-heat-treated sample has the lowest hardness, and the samples sintered at 1090 °C have a larger hardness and reduced elastic moduli. The presented data in Table 1 provide that the effect of the heat-treatment step on the final relative density, much like the effect on grain size, is largely negligible. Increased hardness suggests that densification of the sample results in a strengthening of the matrix while opposing the effect of Hall–Petch. The results obtained here on the relative density are comparable to our previous paper [24] and suggest large densification is possible once the solidus temperature is achieved, and a corresponding increase in relative density resulting from partial super-solidus liquid phase sintering was reflected in the reduced elastic modulus increase. The following discussion provides a quantitative approach to the mechanical properties' relationship to densification.

Gibson and Ashby [35] predict the following relationship for the elastic modulus provided the porosity is homogenous:

$$\frac{E^*}{E_s} = k \left(\frac{\rho^*}{\rho_s}\right)^2 \quad (1)$$

Here, E^* and E_S are the foam (as-printed porous material sample) elastic modulus and bulk (theoretical 100% relative density) elastic modulus, and ρ^* and ρ_S are the foam density and bulk density, respectively. Provided an indentation procedure follows Hertzian contact mechanics, the reduced modulus can be determined as a function of the specimen and indenter properties [36]. Provided below is a fundamental equation that elucidates the elastic moduli through material contact (for example, between an indenter and specimen):

$$\frac{1}{E_r} = \frac{1-\nu^2}{E} + \frac{1-\nu_i^2}{E_i} \tag{2}$$

where E, E_i, and E_r are the specimen elastic modulus, indenter elastic modulus, and reduced elastic modulus, respectively, and ν and ν_i are the Poisson's ratio of the specimen and indenter, respectively. These two equations provide that reduced elastic modulus (and correspondingly, the foam and bulk elastic modulus), is a function of porosity. An example of this relationship was observed in a study by Chen et al. [37], where nanoindentation was applied to porous lanthanum-based perovskite-structured thin films sintered at various temperatures and elastic modulus and hardness were reported. It was shown here that increased porosity could decrease the elastic modulus exceeding an order of magnitude. Using the commonly assigned Poisson's ratio of 0.3 [38] and the literature values for the Berkovich indenter properties [39], the foam elastic moduli for the four different samples were determined. Furthermore, utilizing Equation (1) and assuming $k = 1$, the bulk elastic modulus was calculated and is provided in Table 1 for each sample. The literature does not provide data for the elastic modulus of bulk $Ni_{50}Mn_{30}Ga_{20}$ in the 14M state despite the discussion that the mechanical properties are highly sensitive to alloying additions and compositional variance. However, our values are comparable to Kart and Cagın, who utilized first-principle calculations to determine the elastic modulus of 5M stoichiometric Ni–Mn–Ga to be 144 GPa at 0 K (7M is omitted in this work due to the large computational requirements) [40]. Further work by Kustov et al. [41] reported a significant softening (five to ten times lower) of the elastic modulus values depending on crystallographic direction and temperature for Ni–Mn–Ga. We note that nanoindentation had been performed in this study on mechanically ground, polished, and etched samples, through which minor surface stress and roughness likely accumulated. It should also be noted that the samples in this study have not been previously mechanically trained (see [21] for more details regarding the training process), failing to reduce the twinning stress in this study and reducing twin-boundary motion compliancy.

Finally, SEM imaging was used following nanoindentation to view indents and observe possible twin variant build-up at the indented boundary. In all micrographs except one, variant build-up was absent, suggesting that twin variants may have achieved motion underneath the sample interface or that twin variant motion was not present entirely (see Supplementary Figure S3 as exemplary SEM micrographs of the indent (Figure S3a) with twins visible at the indent and (Figure S3b) without twins). This suggests that twin variant reorientation through the movement of twin boundaries was largely negligible during nanoindentation and the corresponding results. This is also supported by the absence of significant mechanical pop-in behavior (discussed further below). The capability for twin-boundary motion upon nano-load introduction and its impact on mechanical properties is not fully elucidated in this study but it is assumed either minor or negligible. Chmielus et al. [42] discussed surface mechanics and twin-boundary motion in Ni–Mn–Ga single crystals, where it was mentioned that mechanical polishing partially negates the surface hardening on cut samples and reduces twinning stress. However, the residual surface roughness and internal stresses may have prevented twin-boundary motion by increasing the twinning stress to be large enough to resist movement under the nano-load. Furthermore, the literature has not provided that Equations (1) and (2) are compatible but has shown the opportunity for future studies to verify the solid elastic behavior of Ni–Mn–Ga materials, which are of particular importance, especially in thin-film and micro-cantilever designs.

Previous work by Jayaraman et al. [43] discussed the variation in the reduced elastic modulus during nanoindentation depending on the crystallographic orientation in the high-temperature austenite phase. This study assumes randomly oriented grains normal to the nanoindentation surface, leading to varying values of reduced modulus of the mechanically anisotropic martensite unit cell. Although there is no literature report (to our knowledge) on mechanical anisotropy of 14M Ni–Mn–Ga martensite, we predict differences in mechanical properties depending on the crystallographic orientation to the sample interface. The 10 indents per sample were performed at sufficiently large distances (>100 μm) so that each indent was performed on a unique grain. We additionally note that Ni–Mn–Ga material will also exhibit bulk orientation-dependent properties depending on processing history (i.e., polycrystalline, crystalline, textured, etc.). In general, polycrystalline Ni–Mn–Ga is the easiest to manufacture but as stated earlier, the introduction of porosity can improve the highly diminished MFIS compared to its single crystalline counterpart. Mechanical anisotropy in 14M martensitic Ni–Mn–Ga definitely may be of interest in future studies, in particular in parallel with synchrotron grain orientation mapping, but for now provides that bulk elastic properties may be ascertained from indentation.

Clearly, microindentation tests (such as the Vickers test or the Rockwell test) present significant deformation in the microstructure compared to the dimensions of the porous network, increasing the chance that a pore plays a role in the indent. However, pore/matrix interplay still cannot be neglected in nanoindentation. The force–displacement curve example provided in Figure 5 suggests that, on average, the maximum indentation depth is 0.4 μm. However, it is still possible that the plastic zone could impinge upon a free surface under the polished interface, as the plastic zone will be large compared to the indent depth, leading to another source of nanoindentation depth variation [44].

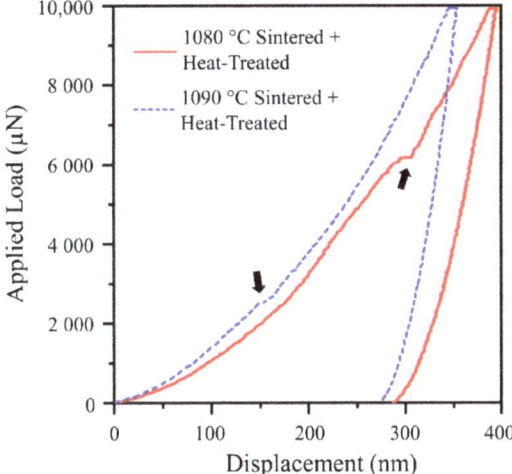

Figure 5. Force–displacement curves extracted from nanoindentation tests on the 1080 °C and 1090 °C sintered and heat-treated samples to display exemplary pop-in observations (see arrows). Larger deformation in the 1080 °C sintered + heat-treated sample corresponds to the smaller hardness value observed between the two samples.

Force–displacement curves obtained during nanoindentation contain various loading points accompanied by brief "bursts" of displacement, referred to as "pop-ins". A pop-in manifests as a horizontal or near-horizontal portion of the force–displacement curve. This phenomenon is commonly accepted to be due to the homogenous nucleation of mobile dislocations at the sample interface [45]. In a study by Aaltio et al. [46], the surface preparation was shown to affect the presence of pop-ins. The samples in this study likely accumulated dislocations during the polishing step. Additionally, plane orientation may have contributed to the presence of pop-ins, that is, depending on how the c-axis of the martensite phase is oriented in the plane of view [47]. As stated previously, this study assumes the indents were performed on random crystal orientations. It is possible that pop-in events contributed to indentation depth variation.

Figure 6 presents DSC curves for the various samples under investigation in this study. Between the 1080 °C heat-treated and non-heat-treated samples, less than a 1 °C difference in average transformation temperature was observed with no change in the Curie temperature (see Table 1 for a summary of transformation temperatures). The 1090 °C sintered and heat-treated sample shows a minor decrease in transformation temperatures (most notably T_{trans}) compared to both 1080 °C sintered samples. However, the largest difference was observed for the 1090 °C non-heat-treated sample, where a 10 °C difference was seen for T_{trans} versus the 1080 °C non-heat-treated sample. Tian et al. [48] discussed that decreasing grain boundary and pore boundary area (through the introduction of an annealing step, for example) will decrease the martensitic phase transformation temperature. However, this does not explain the major difference between the 1090 °C sintered and heat-treated versus the 1090 °C sintered and non-heat-treated samples, as the grain size is comparable. Instead, the large deviation in T_{trans} for the 1090 °C sintered and non-heat-treated sample in this study is attributed to chemical inhomogeneity, compositional variation, decreased atomic ordering, or elemental Mn evaporation [49]. Previous work by Mostafaei et al. [24] on equivalent samples of the present study discussed the accumulation of Mn at the grain boundaries following sintering, which may reduce martensitic phase transformation temperature due to less effective Mn in the grains of the matrix. The addition of a heat-treatment step improves chemical homogeneity, once again increasing the transformation temperature. Thus, our data suggest that the shift to lower transformation temperatures for 1090 °C sintered samples due in part to chemical inhomogeneity and Mn precipitation may be partially increased through the introduction of a heat-treatment step. This assertion is supported in the work by Schlagel [50] where it was discussed the systematic increase in martensitic transformation temperature across a compositional gradient of increasing relative Mn content. It follows that, with decreased Mn available within the uniform grains, the martensitic phase transformation temperature decreases where the effective Mn content is lower. It is also established in the literature that the Mn vapor pressure in the Ni–Mn–Ga solution is relatively large compared to Ni and Ga [51], which enhances the ability of the vapor transport sintering mechanism at elevated temperatures, another means for Mn precipitation. Consequently, enhanced vapor transport of elemental Mn may have resulted in partial evaporation of the species during the 1090 °C sintering step. It is apparent that heat-treatment may remedy the chemical segregation, but cannot return evaporated Mn to the material, thus partial recovery of T_{trans} is observed instead of full recovery. Utilizing EDS following the DSC measurements, we were able to confirm a minor decrease (on the order of 0.5 at %) in the overall Mn content in the 1090 °C sintered samples versus the 1080 °C sintered samples. One additional note is that the relatively high transformation temperatures would correspond with relatively high twinning stress at ambient temperature [52]. This restricts the ability for MFIS to be demonstrated, as twinning stress must be overcome for twin variant reorientation and twin-boundary movement. Future studies may be concerned with compositional or post-processing adjustment to lower this transformation temperature slightly.

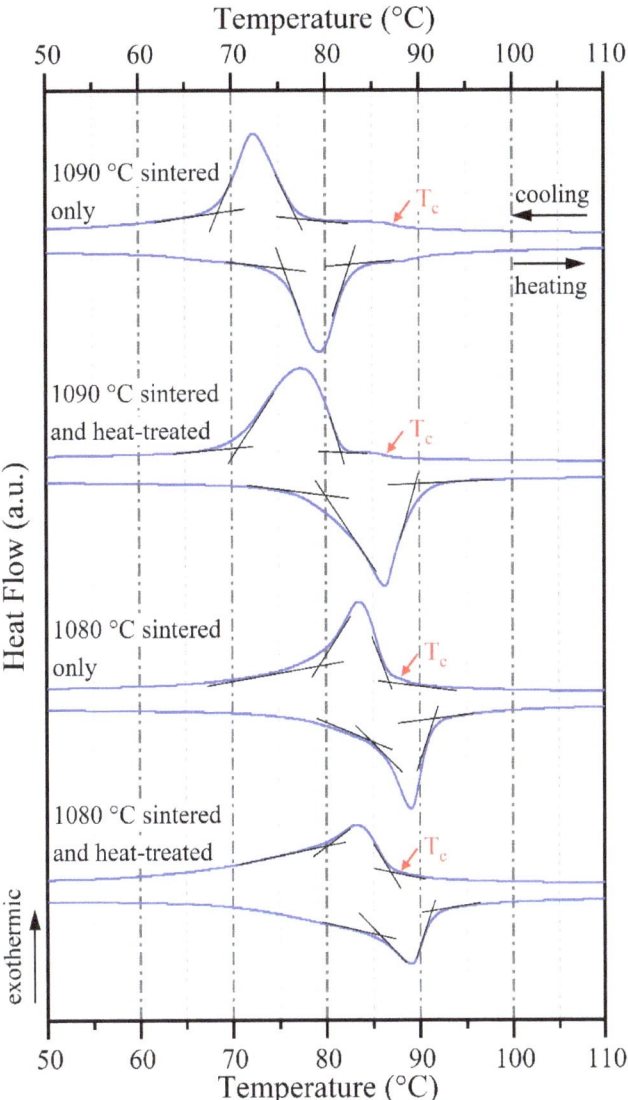

Figure 6. DSC curves presenting the transformation temperatures of the four investigated samples. Tangent lines at transformation onsets and ticks at peak maxima are presented to aid the eye. Curie temperature was determined at heat flow stepwise increase at ~90 °C. See Table 1 for observed transformation temperatures.

Figure 7 presents regions identified as containing many twin boundaries for the 1080 °C and 1090 °C sintered and non-heat-treated electropolished samples and the corresponding AFM and MFM measurements taken at these regions. Based on the presented data, it is concluded that MFM contrast is present in the form of magnetic anisotropy at the interface. In both samples, tightly spaced variants ≤10 μm (many approximately 1–2 μm) in width are observed and exhibit magnetic anisotropy. Laitinen et al. [53] showed similar narrow and finely spaced twin variants revealed by MFM in 14M martensite printed via laser powder bed fusion. In addition, it was shown that MFM contrast was greatly improved after homogenization at 1080 °C for 24 h compared to the as-built material in [52]. Thus, the weaker contrast in the present research may point to a not-fully relaxed structure that could be relieved with additional annealing. This MFM experiment confirms that twins are seen in both 1080 °C and 1090 °C sintered samples with similar magnetic anisotropy behavior, suggesting that MFIS is possible no matter if partial melting is achieved during sintering.

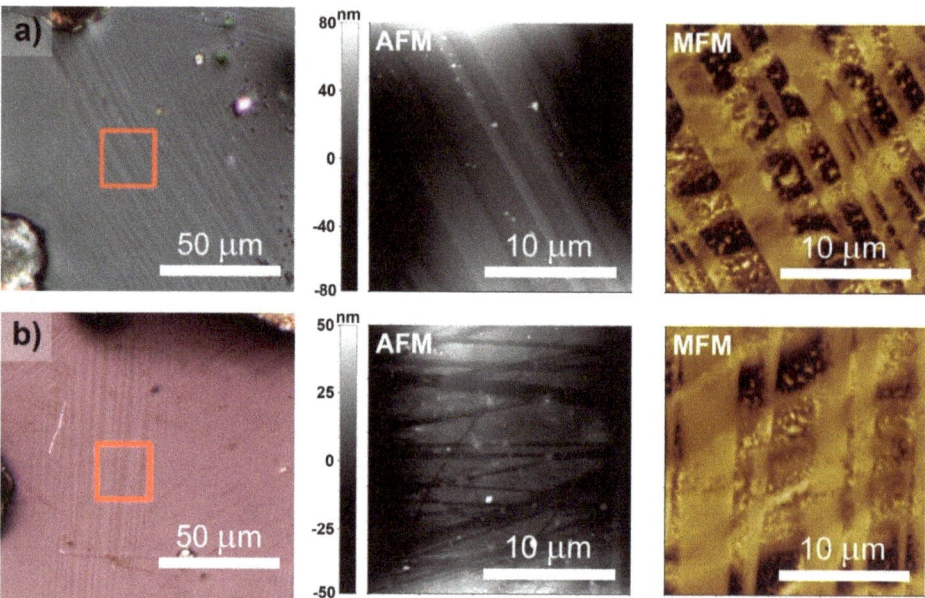

Figure 7. Optical micrographs (left) and corresponding AFM (middle)/MFM (right) scans of the red squares of the left micrographs for the 1080 °C (row (**a**)) and 1090 °C (row (**b**)) sintered samples at twin-dense regions. The square red regions are 20 µm × 20 µm as indicated in the AFM/MFM micrographs.

Saturation magnetization per unit weight (M_{sat} in Table 1) is the observed magnetization at an applied field of 1.5 T (M–H curves in Figure 8). There is a minor increase in saturation magnetization in the 1080 °C sintered sample following heat treatment and nearly identical saturation magnetization in both 1090 °C sintered samples. Additionally, magnetic coercivity (determined using the calculation technique in [54]) approximately 40 mT is observed for the non-heat-treated samples, whereas the heat treatment in both the 1080 °C and 1090 °C sintering temperature cases tightens the hysteresis and reduces magnetic coercivity to approximately 10 mT. All values presented here are comparable to our previously published paper on identical non-heat-treated samples within the bounds of uncertainty [24]. Comparing values in both this and our previous study [24], it is determined that the saturation magnetization is largely indifferent to the heat-treatment step and the magnetic hysteresis tightening exhibited by heat-treated samples likely results from increased chemical ordering and homogenization (including integration of Mn from the grain boundaries to the grains) within the matrix.

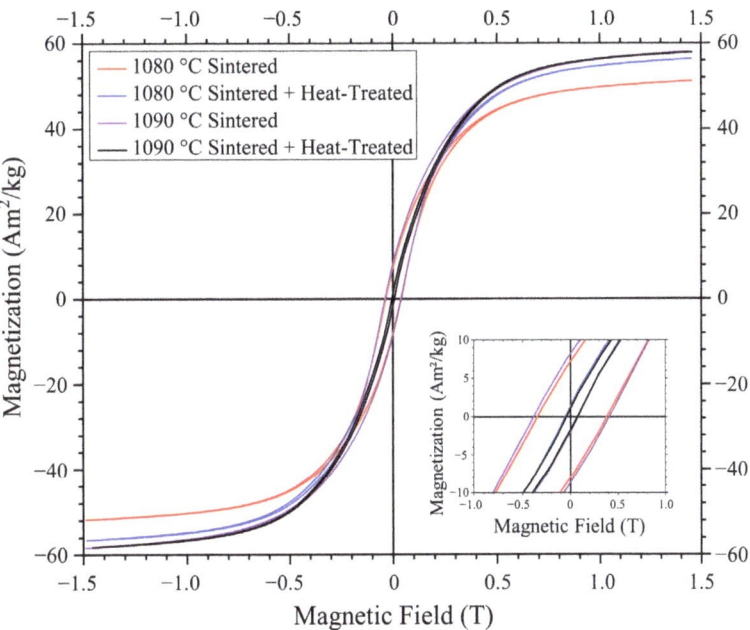

Figure 8. Magnetization (M–H) curves comparing saturation magnetization and coercivity for the different sintering and post-processing conditions. Inset is a magnified view of the hysteresis (note the unit change on the horizontal axis).

4. Conclusions

The sintering of binder-jet 3D printed 14M $Ni_{50}Mn_{30}Ga_{20}$ samples for 2 h at 1080 °C and 1090 °C, respectively, shows a large difference in grain size as identified by neutron diffraction area detector raw data inspection and quantitative stereology. This is accompanied by an increase in relative density (~80% to ~95%) through a 10 °C sintering temperature increase. This minor increase in sintering temperature accompanied a significant difference in mechanical properties, including an increase in hardness and reduced foam elastic moduli, and in transformation temperatures including a decrease in martensitic phase transformation onset. An additional heat-treatment step (annealing and ordering) was seen to only slightly affect the grain size, densification, and mechanical properties. However, heat treatment slightly homogenizes the sample following 1090 °C sintering, evident in the martensitic phase transformation onset temperature change versus the non-heat-treated sample. Magnetic anisotropy in differently sintered samples with twin variants finely spaced in the order of ≤10 μm in width suggests the possibility of MFIS in all presented sintering conditions. Magnetization curves suggested only slight saturation magnetization differences in the samples and hysteresis tightening, due in part to compositional homogenization. This study provided an avenue to identify the solid bulk elastic modulus utilizing equations from porous systems theory, as well as the elastic moduli for samples at respective processing conditions and sintering history. The transformation temperatures of $Ni_{50}Mn_{30}Ga_{20}$ presented in this study provide the capability for further material composition and property tailoring through the introduction of heat-treatment steps and smart alloy design. Finally, by combining tailored composition, properties, grain and pore microstructure, binder-jet 3D printed polycrystalline Ni–Mn–Ga with low blocking stresses and, thus, a significant magnetic field-induced strain, may be achievable.

Supplementary Materials: The following supporting information can be downloaded at: https://www.mdpi.com/article/10.3390/met12050724/s1, Figure S1: SEM micrographs of the (a) 1080 °C sintered and (b) 1090 °C sintered only samples showing twinning, grain boundaries, and inherent porosity; Figure S2: Example grain boundary tracing for grain size determination; Figure S3: Exemplary SEM micrographs of nanoindentations of the 1080 °C sintered only sample (a) with and (b) without twin boundary build-up.

Author Contributions: Conceptualization: A.A. and M.C.; methodology: A.A., J.T. and A.M.; software: A.A., M.B. and R.C.W.; formal analysis: A.A., K.K. and M.C.; investigation: A.A., J.T., M.C., A.S., K.K., V.L. and A.M.; data curation: A.A., K.K., V.L., A.S., A.M., M.B. and R.C.W.; supervision: M.B., R.C.W., M.C. and K.U.; resources: M.C., K.U., M.B. and R.C.W.; writing—original draft preparation: A.A.; writing—review and editing: A.A., M.C., A.M., V.L., J.T., K.K., M.B., R.C.W., A.S. and K.U.; visualization: A.A., A.S. and K.K.; project administration: M.C.; funding acquisition: M.C. and K.U.; All authors have read and agreed to the published version of the manuscript.

Funding: This research was funded by the National Science Foundation (NSF), grant number 1727676, and the Academy of Finland, grant number 325910. J.T. and M.C. would like to acknowledge the Hewlett International Grant Program for travel funding.

Data Availability Statement: Please contact the corresponding investigator for the code utilized for characterization.

Acknowledgments: A.A. would like to acknowledge K. Shields and E. Stevens for DIC micrograph assistance, R. Rodriguez de Vecchis for sample preparation assistance, E. Stevens and J. Schneiderman for additional proofreading, and the Materials Metrology, Characterization, and Learning (MMCL) Lab at the University of Pittsburgh for the opportunity to perform characterization on the investigated samples.

Conflicts of Interest: The authors declare no conflict of interest. Additionally, the funders had no role in the design of the study; in the collection, analysis, or interpretation of data; in the writing of the manuscript, or in the decision to publish the results.

References

1. Ullakko, K.; Huang, J.K.; Kantner, C.; O'Handley, R.C.; Kokorin, V.V. Large magnetic-field-induced strains in Ni_2MnGa single crystals. *Appl. Phys. Lett.* **1996**, *69*, 1966–1968. [CrossRef]
2. Ullakko, K. Magnetically controlled shape memory alloys: A new class of actuator materials. *J. Mater. Eng. Perform.* **1996**, *5*, 405–409. [CrossRef]
3. Henry, C.P.; Bono, D.; Feuchtwanger, J.; Allen, S.M.; O'Handley, R.C. AC field-induced actuation of single crystal Ni–Mn–Ga. *J. Appl. Phys.* **2002**, *91*, 7810. [CrossRef]
4. Toman, J.; Müllner, P.; Chmielus, M. Properties of as-deposited and heat-treated Ni-Mn-Ga magnetic shape memory alloy processed by directed energy deposition. *J. Alloys Compd.* **2018**, *752*, 455–463. [CrossRef]
5. Müllner, P.; Chernenko, V.A.; Kostorz, G. Large cyclic magnetic-field-induced deformation in orthorhombic (14M) Ni–Mn–Ga martensite. *J. Appl. Phys.* **2004**, *95*, 1531–1536. [CrossRef]
6. Tao, T.; Liang, Y.C. Bio-inspired Actuating System for Swimming Using Shape Memory Alloy Composites. *Int. J. Autom. Comput.* **2006**, *3*, 366–373. [CrossRef]
7. Smith, A.R.; Tellinen, J.; Ullakko, K. Rapid actuation and response of Ni–Mn–Ga to magnetic-field-induced stress. *Acta Mater.* **2014**, *80*, 373–379. [CrossRef]
8. Suorsa, I.; Pagounis, E.; Ullakko, K. Magnetic shape memory actuator performance. *J. Magn. Magn. Mater.* **2004**, *272–276*, 2029–2030. [CrossRef]
9. Musiienko, D.; Saren, A.; Straka, L.; Vronka, M.; Kopeček, J.; Heczko, O.; Sozinov, A.; Ullakko, K. Ultrafast actuation of Ni-Mn-Ga micropillars by pulsed magnetic field. *Scr. Mater.* **2019**, *162*, 482–485. [CrossRef]
10. Witherspoon, C.; Zheng, P.; Chmielus, M.; Dunand, D.C.; Müllner, P. Effect of porosity on the magneto-mechanical behavior of polycrystalline magnetic shape-memory Ni–Mn–Ga foams. *Acta Mater.* **2015**, *92*, 64–71. [CrossRef]
11. Sozinov, A.; Likhachev, A.A.; Lanska, N.; Ullakko, K. Giant magnetic-field-induced strain in NiMnGa seven-layered martensitic phase. *Appl. Phys. Lett.* **2002**, *80*, 1746. [CrossRef]
12. Roy, S.; Blackburn, E.; Valvidares, S.M.; Fitzsimmons, M.R.; Vogel, S.C.; Khan, M.; Dubenko, I.; Stadler, S.; Ali, N.; Sinha, S.K.; et al. Delocalization and hybridization enhance the magnetocaloric effect in Cu-doped Ni_2MnGa. *Phys. Rev. B* **2009**, *79*, 235127. [CrossRef]
13. Wilson, S.A.; Jourdain, R.P.J.; Zhang, Q.; Dorey, R.A.; Bowen, C.R.; Willander, M.; Wahab, Q.U.; Willander, M.; Al-hilli, S.M.; Nur, O.; et al. New materials for micro-scale sensors and actuators: An engineering review. *Mater. Sci. Eng. R Rep.* **2007**, *56*, 1–129. [CrossRef]
14. Lázpita, P.; Rojo, G.; Gutiérrez, J.; Barandiaran, J.M.; O'Handley, R.C. Correlation between magnetization and deformation in a NiMnGa shape memory alloy polycrystalline ribbon. *Sens. Lett.* **2007**, *5*, 65–68. [CrossRef]

15. Caputo, M.; Solomon, C.V.; Nguyen, P.-K.; Berkowitz, A.E. Electron Microscopy Investigation of Binder Saturation and Microstructural Defects in Functional Parts Made by Additive Manufacturing. *Microsc. Microanal.* **2016**, *22*, 1770–1771. [CrossRef]
16. Ullakko, K.; Chmielus, M.; Müllner, P. Stabilizing a fine twin structure in Ni-Mn-Ga samples by coatings and ion implantation. *Scr. Mater.* **2015**, *94*, 40–43. [CrossRef]
17. Chmielus, M.; Witherspoon, C.; Ullakko, K.; Müllner, P.; Schneider, R. Effects of Surface Damage on Twinning Stress and the Stability of Twin Microstructures of Magnetic Shape Memory Alloys. *Acta Mater.* **2011**, *59*, 2948–2956. [CrossRef]
18. Chmielus, M.; Glavatskyy, I.; Hoffmann, J.U.; Chernenko, V.A.; Schneider, R.; Müllner, P. Influence of Constraints and Twinning Stress on Magnetic Field-Induced Strain of Magnetic Shape-Memory Alloys. *Scr. Mater.* **2011**, *64*, 888–891. [CrossRef]
19. Caputo, M.P.; Solomon, C.V. A Facile Method for Producing Porous Parts with Complex Geometries from Ferromagnetic Ni-Mn-Ga Shape Memory Alloys. *Mater. Lett.* **2017**, *200*, 87–89. [CrossRef]
20. Boonyongmaneerat, Y.; Chmielus, M.; Dunand, D.C.; Müllner, P. Increasing Magnetoplasticity in Polycrystalline Ni-Mn-Ga by Reducing Internal Constraints through Porosity. *Phys. Rev. Lett.* **2007**, *99*, 247201. [CrossRef]
21. Chmielus, M.; Zhang, X.X.; Witherspoon, C.; Dunand, D.C.; Müllner, P. Giant Magnetic-Field-Induced Strains in Polycrystalline Ni–Mn–Ga Foams. *Nat. Mater.* **2009**, *8*, 863–866. [CrossRef] [PubMed]
22. Lu, K.; Reynolds, W.T. 3DP Process for Fine Mesh Structure Printing. *Powder Technol.* **2008**, *187*, 11–18. [CrossRef]
23. Taylor, S.L.; Shah, R.N.; Dunand, D.C. Ni-Mn-Ga Micro-Trusses via Sintering of 3D-Printed Inks Containing Elemental Powders. *Acta Mater.* **2018**, *143*, 20–29. [CrossRef]
24. Mostafaei, A.; Rodriguez De Vecchis, P.; Stevens, E.L.; Chmielus, M. Sintering Regimes and Resulting Microstructure and Properties of Binder Jet 3D Printed Ni-Mn-Ga Magnetic Shape Memory Alloys. *Acta Mater.* **2018**, *154*, 355–364. [CrossRef]
25. Mostafaei, A.; Kimes, K.A.; Stevens, E.L.; Toman, J.; Krimer, Y.L.; Ullakko, K.; Chmielus, M. Microstructural Evolution and Magnetic Properties of Binder Jet Additive Manufactured Ni-Mn-Ga Magnetic Shape Memory Alloy Foam. *Acta Mater.* **2017**, *131*, 482–490. [CrossRef]
26. Caputo, M.P.; Berkowitz, A.E.; Armstrong, A.; Müllner, P.; Solomon, C.V. 4D Printing of Net Shape Parts Made from Ni-Mn-Ga Magnetic Shape-Memory Alloys. *Addit. Manuf.* **2018**, *21*, 579–588. [CrossRef]
27. Mostafaei, A.; Elliott, A.M.; Barnes, J.E.; Li, F.; Tan, W.; Cramer, C.L.; Nandwana, P.; Chmielus, M. Binder jet 3D printing—Process parameters, materials, properties, modeling, and challenges. *Prog. Mater. Sci.* **2021**, *119*, 100707. [CrossRef]
28. Wimpory, R.C.; Mikula, P.; Šaroun, J.; Poeste, T.; Li, J.; Hofmann, M.; Schneider, R. Efficiency Boost of the Materials Science Diffractometer E3 at BENSC: One Order of Magnitude Due to a Horizontally and Vertically Focusing Monochromator. *Neutron News* **2008**, *19*, 16–19. [CrossRef]
29. Chmielus, M.; Witherspoon, C.; Wimpory, R.C.; Paulke, A.; Hilger, A.; Zhang, X.; Dunand, D.C.; Müllner, P. Magnetic-Field-Induced Recovery Strain in Polycrystalline Ni–Mn–Ga Foam. *J. Appl. Phys.* **2010**, *108*, 123526. [CrossRef]
30. Rueden, C.T.; Schindelin, J.; Hiner, M.C.; DeZonia, B.E.; Walter, A.E.; Arena, E.T.; Eliceiri, K.W. ImageJ2: ImageJ for the next Generation of Scientific Image Data. *BMC Bioinform.* **2017**, *18*, 529. [CrossRef]
31. Schindelin, J.; Arganda-Carreras, I.; Frise, E.; Kaynig, V.; Longair, M.; Pietzsch, T.; Preibisch, S.; Rueden, C.; Saalfeld, S.; Schmid, B.; et al. Fiji: An Open-Source Platform for Biological-Image Analysis. *Nat. Methods* **2012**, *9*, 676–682. [CrossRef] [PubMed]
32. Mostafaei, A.; Rodriguez De Vecchis, P.; Nettleship, I.; Chmielus, M. Effect of Powder Size Distribution on Densification and Microstructural Evolution of Binder-Jet 3D-Printed Alloy 625. *Mater. Des.* **2019**, *162*, 375–383. [CrossRef]
33. McIntyre, G.J. Area Detectors in Single-Crystal Neutron Diffraction. *J. Phys. D Appl. Phys.* **2015**, *48*, 504002. [CrossRef]
34. Overholser, R.W.; Wuttig, M.; Neumann, D.A. Chemical Ordering in Ni-Mn-Ga Heusler Alloys. *Scr. Mater.* **1999**, *40*, 1095–1102. [CrossRef]
35. Gibson, L.J.; Ashby, M.F. *Cellular Solids: Structure and Properties*, 2nd ed.; Cambridge University Press: Cambridge, UK, 1999.
36. Chicot, D.; N'Jock, M.Y.; Roudet, F.; Decoopman, X.; Staia, M.H.; Puchi-Cabrera, E.S. Some Improvements for Determining the Hardness of Homogeneous Materials from the Work-of-Indentation. *Int. J. Mech. Sci.* **2016**, *105*, 279–290. [CrossRef]
37. Chen, Z.; Wang, X.; Bhakhri, V.; Giuliani, F.; Atkinson, A. Nanoindentation of Porous Bulk and Thin Films of $La_{0.6}Sr_{0.4}Co_{0.2}Fe_{0.8}O_3-\delta$. *Acta Mater.* **2013**, *61*, 5720–5734. [CrossRef]
38. Kohl, M.; Agarwal, A.; Chernenko, V.A.; Ohtsuka, M.; Seemann, K. Shape Memory Effect and Magnetostriction in Polycrystalline Ni–Mn–Ga Thin Film Microactuators. *Mater. Sci. Eng. A* **2006**, *438–440*, 940–943. [CrossRef]
39. Zorzi, J.E.; Perottoni, C.A. Estimating Young's Modulus and Poisson's Ratio by Instrumented Indentation Test. *Mater. Sci. Eng. A* **2013**, *574*, 25–30. [CrossRef]
40. Ozdemir Kart, S.; Cagın, T. Elastic Properties of Ni_2MnGa from First-Principles Calculations. *J. Alloys Compd.* **2010**, *508*, 177–183. [CrossRef]
41. Kustov, S.; Saren, A.; Sozinov, A.; Kaminskii, V.; Ullakko, K. Ultrahigh Damping and Young's Modulus Softening Due to a/b Twins in 10M Ni-Mn-Ga Martensite. *Scr. Mater.* **2020**, *178*, 483–488. [CrossRef]
42. Chmielus, M.; Rolfs, K.; Wimpory, R.; Reimers, W.; Müllner, P.; Schneider, R. Effects of Surface Roughness and Training on the Twinning Stress of Ni–Mn–Ga Single Crystals. *Acta Mater.* **2010**, *58*, 3952–3962. [CrossRef]
43. Jayaraman, A.; Kiran, M.S.R.N.; Ramamurty, U. Mechanical Anisotropy in Austenitic NiMnGa Alloy: Nanoindentation Studies. *Crystals* **2017**, *7*, 254. [CrossRef]
44. Kramer, D.; Huang, H.; Kriese, M.; Robach, J.; Nelson, J.; Wright, A.; Bahr, D.; Gerberich, W.W. Yield Strength Predictions from the Plastic Zone around Nanocontacts. *Acta Mater.* **1998**, *47*, 333–343. [CrossRef]

45. Li, T.L.; Gao, Y.F.; Bei, H.; George, E.P. Indentation Schmid Factor and Orientation Dependence of Nanoindentation Pop-in Behavior of NiAl Single Crystals. *J. Mech. Phys. Solids* **2011**, *59*, 1147–1162. [CrossRef]
46. Aaltio, I.; Liu, X.W.; Valden, M.; Lahtonen, K.; Söderberg, O.; Ge, Y.; Hannula, S.-P. Nanoscale Surface Properties of a Ni–Mn–Ga 10M Magnetic Shape Memory Alloy. *J. Alloys Compd.* **2013**, *577*, S367–S371. [CrossRef]
47. Davis, P.H.; Efaw, C.M.; Patten, L.K.; Hollar, C.; Watson, C.S.; Knowlton, W.B.; Müllner, P. Localized Deformation in Ni-Mn-Ga Single Crystals. *J. Appl. Phys.* **2018**, *123*, 215102. [CrossRef]
48. Tian, X.H.; Sui, J.H.; Zhang, X.; Zheng, X.H.; Cai, W. Grain Size Effect on Martensitic Transformation, Mechanical and Magnetic Properties of Ni–Mn–Ga Alloy Fabricated by Spark Plasma Sintering. *J. Alloys Compd.* **2012**, *514*, 210–213. [CrossRef]
49. Qian, M.F.; Zhang, X.X.; Wei, L.S.; Geng, L.; Peng, H.X. Effect of Chemical Ordering Annealing on Martensitic Transformation and Superelasticity in Polycrystalline Ni–Mn–Ga Microwires. *J. Alloys Compd.* **2015**, *645*, 335–343. [CrossRef]
50. Schlagel, D.; Wu, Y.; Zhang, W.; Lograsso, T. Chemical Segregation during Bulk Single Crystal Preparation of Ni–Mn–Ga Ferromagnetic Shape Memory Alloys. *J. Alloys Compd.* **2000**, *312*, 77–85. [CrossRef]
51. Dunand, D.C.; Müllner, P. Size Effects on Magnetic Actuation in Ni-Mn-Ga Shape-Memory Alloys. *Adv. Mater.* **2011**, *23*, 216–232. [CrossRef]
52. Laitinen, V.; Sozinov, A.; Saren, A.; Chmielus, M.; Ullakko, K. Characterization of As-Built and Heat-Treated Ni-Mn-Ga Magnetic Shape Memory Alloy Manufactured via Laser Powder Bed Fusion. *Addit. Manuf.* **2021**, *39*, 101854. [CrossRef]
53. Laitinen, V.; Sozinov, A.; Saren, A.; Salminen, A.; Ullakko, K. Laser Powder Bed Fusion of Ni-Mn-Ga Magnetic Shape Memory Alloy. *Addit. Manuf.* **2019**, *30*, 100891. [CrossRef]
54. Wang, H.; Zhang, H.; Wang, Y.; Tan, W.; Huo, D. Spin glass feature and exchange bias effect in metallic Pt/antiferromagnetic LaMnO3 heterostructure. *J. Phys. Condens. Mat.* **2021**, *33*, 285802. [CrossRef] [PubMed]

MDPI
St. Alban-Anlage 66
4052 Basel
Switzerland
Tel. +41 61 683 77 34
Fax +41 61 302 89 18
www.mdpi.com

Metals Editorial Office
E-mail: metals@mdpi.com
www.mdpi.com/journal/metals

www.ingramcontent.com/pod-product-compliance
Lightning Source LLC
LaVergne TN
LVHW070629100526
838202LV00012B/764